MULTI-PLATFORM DATABASE CONSULTING

DATABASE USEAGE ON MAINFRAME, MID-TIER AND NT SERVERS

by

John V. McLean

authorHOUSE

1663 Liberty Drive, Suite 200
Bloomington, Indiana 47403
(800) 839-8640
www.authorhouse.com

Table of Contents

1.0   Introduction

This book is a "check list" and is specifically designed for DB2 DBAs who are or want to be DB2 DBA Consultants. This book is "focused" towards performance tuning on UNIX and AIX Platforms, but topics involving OS390 and NT are also covered. However, any IT personnel working with DB2 from Company Directors to Application Developers may make valuable use of the information in this book.

In today's fast paced world, answers are needed quickly, at the same time these answers must be accurate and meaningful. This book is aimed at addressing this situation.

The book is designed as a check list, but it is also more than just a list of items to tick off without true understanding of the implications involved. Therefore, effort has been expended on detailed explanations where appropriate.

DB2 is an ever evolving technology, and this book focuses on the trends and features that will keep the reader up to date with 'state of the art' tips, and techniques that makes DB2 the premier database engine of today.

Today, you are worth what you know, consequently, increasing your knowledge base will without doubt increase your net worth. As enterprises rely more and more to succeed on the information they acquire, the importance of the database and sales of database engines such as DB2 are growing exponentially.

Skilled DB2 DBAs are now in demand more than ever before, on a world wide basis. Because of the Internet, boundaries and barriers that were once formidable 'obstacles' to overcome have all but vanished. Therefore, the skilled DBA has virtually the whole world as a market. In addition to this, the price of hardware is always heading downward for the same type of 'box', this contributing to the expansion of DB2 user participants in the market and inevitably increasing demand for DB2 DBAs.

So, if your sights are set on ever improving the service you provide, with speed and accuracy, this check list is there to help you keep your Client satisfied with reference to DB2.

2.0 Consulting

<u>Overview</u>

Not everyone can be a DB2 DBA Consultant! It takes a rare breed. People who are restless, want to succeed and do not mind moving about are potential candidates for Consultancy. DB2 DBAs who are Consultants are relatively rare and very special people!

What do you need to be a Consultant?

1. **Experience.** You need to have a lot of experience, be an expert in what you do (continually strive to be the best), and be able to keep up with the ever evolving technology that you practice. You will be learning continuously. Also, you will be "pulled" in many directions at the same time, the key to success is to stay focused.

2. **Attitude.** Attitude will make or break your efforts at being a successful Consultant. Remember, you are not the boss, the Client is the boss. You are at the Client's 'site' to advise, share your expertise and bring the Project to a successful conclusion. If you fight or argue with the Client, you'll lose, and you're out. So remember, no tempers, if you have a temper control it, otherwise being a Consultant is not for you.

3. **Set yourself up with a Corporation.** It is virtuallyimpossible to operate as an Independent Consultant without having a corporation through which to function. You may set up a corporation yourself, use an accountant, or use an agency like The Company Corporation in Wilmington, Delaware. I have found The Company Corporation (Phone:- 302-636-5440) to be an excellent agent. They will set you up, and for a small fee of around $250/year take care of all your annual paperwork, which is substantial. If you are lucky enough to 'land' an overseas assignment, set up a corporation in the country where you will be working, that way you will not have to wait for work permits, you can start invoicing through your overseas company immediately. If you work in Europe you could set

up a Corporation in every country you contemplate working in; accounting wizards in London will set you up in a few days at reasonable cost. Adapt to the local customs. Look to the future and lay the foundation of having several DBAs working for you in the future.

4. Pay your taxes! "Get" a good accountant (CPA) and a good attorney. Make sure that these two important people live locally, otherwise you may have delays and communication problems in case of an emergency.

5. Make sure you have copies of the following forms ready at all times for any interested party:-

    a. <u>Your Company's Certificate of Incorporation.</u> This document proves that you can operate through your corporation and engage in contracts with other companies on a Corp. to Corp (1099) basis. Remember, you are an employee of your corporation, therefore you pay taxes on a W2 basis to your corporation and are not liable to pay any withholding taxes to your Broker, Client or Recruiter. Your Broker, Client or Recruiter must pay you the gross amount if you are incorporated i.e. Rate * Hours worked (verified by signed Time Sheet).

    b. <u>Certificate of Good Standing.</u> This document basically says that you are who you say you are and that you have filed your State Franchise Taxes (the State in which your Company was incorporated) and that your Annual Reports are up to date.

    c. <u>941 Quarterly Returns.</u>
        When you have a Company of your own you are given an Employer Identification Number (a.k.a. Federal Id. Number). This is a unique number that applies only to your Company. Every month you pay taxes related to Form 941. You will be given a coupon book, and you must pay these taxes before the 16th of

the month following the month they were due. Make the check out to your Bank and deposit the coupon and check at your bank every month before the 16th. At the end of each quarter you will be mailed a 941 Quarterly Returns Form. Complete this form, sign and copy it before you mail it back to the IRS. Keep the signed copy. Ask your Accountant to advise you how to complete Form 941.

d. See your Accountant to advise you on payment of your State and Federal Taxes.

e. Make friends with a good Lawyer, you never know when you may need a Lawyer.

6. You will need some ancillary documentation like:-

1. Business Card (have, at least, Your Name, Company Name, Address, Telephone Number and E-Mail address on your business card).

2. Letter Head (keep it simple!).

3. Business Envelope.

7. Resume – Always keep your Resume up to date. You never know when someone will ask you for the most recent copy of your resume. Keep your resume on a floppy disk in your Briefcase or on your Laptop.

8. Create your own Web Site and publish your skills and make sure you respond to every enquiry. Meet or exceed the criteria you have set for yourself and your Company.

9. Define a Rate Table by location, duration and skill level. Local Brokers/Agents will be able to help you establish a reasonable approach to rating. Do not ever try to 'rip off' your Broker or Client. They will find out and get rid of you at the first opportune moment.

10.   At interview – know your stuff, be direct, be professional and look the part. Always wear a dark suit and sober tie. Your shoes must gleam! Do not try to bamboozle the interviewer with jargon and buzz words, they will soon get tired of you. Rather, show great interest in the Project, ask a lot of relevant questions. Smile from time to time, show the Client you're human! "Connect" with the interviewer.

11.   Last but not least remember K.I.S.S. (Keep it Simple Stupid). Clients hate a smart Alec. Use short meaningful sentences. Simple words. Crystal clear ideas. You do want to succeed as a Consultant, don't you?? Then focus completely on what you are doing!!

3.0   DB2 General

This book is divided into three parts to cover the three basic platforms defined by their operating systems i.e. OS/390, UNIX/AIX and NT. At the time of writing there is not absolute transparency between these three platforms, therefore they have to be addressed separately. However, the ever evolving release definitions of UDB will ultimately lead to a platformless version of the product, at which time an upgraded version of this book will be available. DB2 could one day become invisible to the user so this book lets us briefly review the three separate platforms. Consequently, the knowledge contained in this book will only increase the demand for your services!

briefly:-

OS/390   - This platform relates to DB2 UDB on Mainframe Computers known collectively as S/390 or System 390 or zOS. The OS/390 relates to the operating system which is the successor to the MVS-ESA operating system. The concept of expensive "big iron" chugging away by itself 'lifting' heavy batch background loads and looked upon as dispensable is now very dated.

With the advent of CMOS and other key innovative upgrades by IBM (zOS), the S/390 has become a key player in the hardware configuration of many companies. In today's world, the S/390 provides huge benefits when used as a high end server that can run together with several other servers including AIX (UNIX) and NT in a distributed environment. Severs like the S/390 supply high-volume transaction processing capability, and may be utilized for such enterprising ventures as data warehousing and the building of data marts.

The modern S/390 servers may be configured to share memory and DASD with other similar machines and can be set up so that they can be connected together to form a 'Parallel Sysplex' (Parallel Processing Systems Complex) to share work loads, data and processing capability. So "Big Iron" as it was once known (still is in some circles) is not dead, but very much alive and in it's own way, "leading the way"!

AIX

AIX has undergone huge changes over the last few years. The hardware has expanded to include multi-processor architecture and Power PCs. A very wide range of operating system services have now to be catered for, including clustering, unlimited upward scalability and new protocols.

SMP (Symmetric Multi Processor) nodes of a clustered environment are connected by fiber optic channels and dynamic switching devices. This set up enables parallel query processing against warehouses and also individual maintenance of each node without having to take the whole system down, just the node in question. The Parallel configuration is also scalable (how deep are your Client's pockets?) which enables the addition of Processing power and DASD as the demand for Resources increases.

NT

NT enables you to use DB2 UDB on your PC. Your PC may be used as a server or a client. You can be a client and connect to remote servers. Your PC must be of adequate size and capacity to run DB2 UDB, this may be ascertained by reading the Accompanying documentation or by accessing the IBM Work site.

Future

The future for DB2 looks very promising, and so also the future for DB2 DBAs. As the price of hardware continues to drop and numerous bench mark tests identify DB2 as the fastest and still the cheapest RDBMS on the market, with widest appeal across continents, people that can administer DB2 to efficiently retrieve data quickly are becoming more and more sought after in the market. Not everyone can be a DB2 DBA. It takes a special type of person to make a good DBA. Not every programmer or analyst likes DBA type of work, leave alone having the ability of successfully carrying out this type of work. But for those who start and can stay the course, and are good at it, the future holds great promise with limitless possibilities. Remember, there are "No Limits!!!"

4.0 OS/390 (Mainframe - Platform)

4.0.1 SMP/E (System Modification Program/Extended)

SMP/E helps you install DB2 on the OS/390 platform.

IBM sends you tapes or cartridges containing DB2, and your first task is to load the data sets on the tapes or cartridges into DB2 Libraries.

To load these DB2 libraries use SMP/E. SMP/E processes the installation tapes or cartridges and creates DB2 Distribution Libraries, DB2 Target Libraries and SMP/E control data sets.

DB2 provides many jobs that invoke SMP/E. These jobs are stored on one of the tapes or cartridges you receive.

You should receive six or more standard 9-track magnetic tapes recorded at 6250 BPI and six 3480 cartridges, or six 4mm cartridges, these combinations depend on the features you ordered. For more information reference the DB2 Installer User's Guide.

The tapes or cartridges you receive from IBM will be in SMP RELFILE format. The first file contains the SMP/E modification control statements in RELFILE format and the subsequent files contain IEBCOPY unloaded partitioned data sets for SMP/E to process.

Each tape or cartridge supplied by IBM has one or more FMIDs (modification identifiers) that SMP/E uses to distinguish the separate parts of DB2. For example: IRLM is distributed with both IMS and DB2 and therefore has separate FMIDs.

In addition to these tapes and cartridges you receive a set of documents including the "IBM Database 2 Program Directory". You should read this document before even thinking about making any installation moves. This document identifies the contents of the FMIDs.

If you plan to use CLI (DB2 Call Level Interface) there are additional jobs you need to run. Reference the Call Level Interface Guide before proceeding.

Before installing DB2 you must use the IBM ServiceLink facility to check for PSP updates to the documentation.

4.0.2 Data Sharing

Data Sharing is a reality that provides greater access to data. Data Sharing gives the user expanded capacity so instead of a single computer system (CPC) multiple CPCs can be used to access data. Users are divided into data sharing groups. DB2 uses a coupling facility to provide communications between members of a data sharing group.

Consider more than one Coupling Facility. Thus, if a coupling facility fails you can always switch to another.

4.0.2.1 Data Sharing Group Bufferpool Duplexing

1. Activate a CFRM (Coupling Facility Resource Management) policy specifying <u>DUPLEX (ENABLE)</u>.

2. SETXCF rebuilds the Coupling Facility:- Start Rebuild Duplex STRNAME=strname (Structure Name).

3. Implement Group Buffer Pool Duplex with more than one SYSPLEX Timer. Do not use **UNDESIG** as the Group Name

To share data, the DB2 Subsystem must belong to the Data Sharing Group.

A Data Sharing group is an OS/390 XCF (Cross System Coupling Facility)

Data Sharing Group requires a Sysplex Environment.

The coupling facility is the component that manages the shared resources of the connected CPCs (Computer Processing Components).

DB2 utilizes the Coupling Facility to provide data sharing groups with coordinated locking, bufferpools, and communication. MVS Version 5 and above are required to install a DB2 coupling facility.

A minimum of one Sysplex Timer must be operational. The Sysplex Timer keeps the processor timestamps synchronized for all DB2s instances in the data sharing group.

Connection to shared DASD - The user data, system catalog, directory data and MVS catalog data must all reside on shared DASD.

The DB2 Logs and boot strap data sets (BSDS) belong to each DB2 member individually but they too must reside on shared DASD.

The installation must have a security system that supports security in a Parallel Sysplex environment before implementing DB2 data sharing. RACF Version 2 Release 1
and higher provides this capability.

Shared Objects:- DB2 Catalog, DB2 Directory, Coupling Facility Structures, Lock Structures, Group Bufferpools, SCA (Shared Communication Area).
Non-Shared Objects:- Archive and Active Logs, BSDS
                    (DSNDB07), Sort, RID and EDM
                    Pools, Local Bufferpools, Trace Data

5.0   DB2 - Mainframe Utilities

5.0.1 CHECK DATA

The CHECK utility checks the integrity of DB2 Data structures. It checks referential integrity between two tables, It ensures data values conforms to check constraints. It also checks DB2 indexes for consistency.

<u>Sample JCL</u>

```
//DB2JOB1  JOB (UTILITY),'DB2 CHECK DATA',MSGCLASS=X,CLASS=X,
//             NOTIFY=USER
//*
//* CHECK DATA UTILITY
//*
//UTIL EXEC DSNUPROC,SYSTEM=DSN,UID='CHECKDATA',UTPROC=''
//*
//* WORK DATA SETS
//*
//DSNUPROC.SORTWK01  DD  UNIT=SYSDA,SPACE=(CYL,(5,1))
//DSNUPROC.SORTWK02  DD  UNIT=SYSDA,SAPCE=(CYL,(5,1))
//DSNUPROC.SORTOUT   DD   DSN=&&SORTOUT,
//             UNIT=SYSDA,SPACE=(CYL,(1,1))
//DSNUPORC.SYSERR       DD   DSN=&&SYSERR,
//             UNIT=SYSDA,SPACE=(CYL,(1,1))
//DSNUPROC.SYSUT1      DD   DSN=&&SYSUT1,
//             UNTI=SYSDA,SPACE=(CYL,(1,1))
//DSNUPROC.UTPRINT    DD   SYSOUT=X
//*
//* UTILITY INPUT CONTROL CARDS. DELETE ALL BAD ROWS AND PUT
//* THEM IN AN EXCEPTION TABLE
//DSNUPROC.SYSIN  DD  *
     CHECK DATA TABLESPACE DBNAME1.TSNAME1
     FOR EXCEPTION IN TBNAME1.ONE
       USE TBNAME1.ONE_EXCEPT
     SCOPE ALL  DELETE YES
/*
//*****************************************************************************
//
```

TBNAME1.ONE_EXCEPT is the table where the errors go, create it exactly like the prime table:-

CREATE TABLE TBNAME1.ONE_EXCEPT LIKE TBNAME1.ONE

5.0.2 DB2 COPY

```
//DB2JOB1  JOB (UILITY),'FULL IMAGE COPY', CLASS=X,MSGCLASS=X,
//              NOTIFY=USER
//*
//*  DB2 FULL COPY
//*
//COPY  EXEC DSNUPROC,SYSTEM=DSN,UID='USERID',UTPROC=''
//*
//DSNUPROC.SYSCOPY1 DD DSN=CAT.FULLCOPY.SEQ.DATASET1(+1),
//              DISP=(MOD,CATLG),DCB=SYS1.MODEL,
//              SPACE=(CYL,(6,2),RLSE),UNIT=3390
//DSNUPROC.SYSCOPY2 DD DSN=CAT.FULLCOPY.SEQ.DATASET2(+1),
//              DISP=(MOD,CATLG),DCB=SYS1.MODEL,
//              SPACE=(CYL,(6,2)RLSE),UNIT=3390
//DSNUPROC.SYSIN         DD     *
COPY  TABLESPACE DBNAME1.TSNAME1
        COYDDN (SYSCOPY1, SYSCOPY2)
        SHRLEVEL REFERENCE
        DSNUM ALL FULL YES
//*
//
//*********************************************************************
```

5.0.3 Copy Incrementally

```
//DB2JOB1 JOB (UTILITY), 'INCREMENTAL COPY', CLASS=X, MSGCLASS=X,
//              NOTIFY=USER
//*
//* DB2 – INCREMENTAL COPY
//*
//COPY EXEC DSNUPROC,SYSTEM=DSN,UID'USERID',ITPROC''
//*
//DSNUPORC.SYSCOPY DD DSN=CAT.INCRCOPY.SEQ.DATASET(+1),
//              DISP=(MOD,CATLG),DCB=SYS1.MODEL,
//              SPACE=(CYL,(5,2)RLSE),UNIT=3390
//DSNUPROC.SYSIN    DD *
        COPY TABLESPACE DBNAME1.TSNAME. SHRLEVEL REFERENCE
            DNSUM ALL  FULL NO
/*
//*********************************************************************
```

5.0.4 DB2 INDEX COPY

```
//DB2JOB1  JOB  (UTILITY), 'INDEX COPY', CLASS=X, MSGCLASS=X,
//               NOTIFY=USER
//*
//* DB2 INDEX COPY
//DSNUPROC.SYSIN  DD  *
       COPY INDEXSPACE DBNAME1.INDXSPN1
       SHRLEVEL REFERENCE
//
```

5.0.4.1 DSNTEP2

DSNTEP2 may be used for executing a variety of SQL Statements. An example of a JCL set
to execute this utlitity is given below. The PLAN name and load libraries will vary
from one installation to the next.

```
//M1DXTEP2  JOB ('BHGBHG60710',B106),'MCLEAN',REGION=8M,CLASS=A,
//        MSGCLASS=Q,NOTIFY=M1DX
//*****************************************************************
//**
//** DB2 SAMPLE DSNTEP2 JCL
//**
//*****************************************************************
//JOBLIB   DD DSN=HCFA1.@DB2.P1.DB2I.SDSNLOAD,DISP=SHR
//BATCHSQL EXEC PGM=IKJEFT01,DYNAMNBR=20
//SYSTSPRT DD SYSOUT=*
//SYSPRINT DD SYSOUT=*
//SYSUDUMP DD SYSOUT=*
//SYSTSIN  DD *
DSN SYSTEM(DB2T)
RUN PROGRAM(DSNTEP2) PLAN(DSNTEP71) -
LIB('M1DX.@DB2.P1.DB2I.RUNLIB.LOAD')
END
/*
//SYSIN   DD  *
SELECT COUNT(*) FROM SYSIBM.SYSTABLES;

SELECT * FROM DBA$MCS2.MCS_PAYMENT_CYCLE;
/*
//
```

John V. McLean

5.0.5 DB2 OS390 Load Utility

The LOAD utility is used to accomplish bulk inserts to DB2 Tables. It can add rows
to a table, retaining the current data, or it can replace existing rows with new data.

```
//DB2JOBU  JOB (UILITY),'LOAD', CLASS=X,MSGCLASS=X,
//                 NOTIFY=USER
//*
//*   DB2 LOAD UTILITY
//*
//*
//UTIL EXEC  DSNUPROC,SYSTEM=DSN,UID'USERID',UTPORC=''
//*
//* WORK DATASETS
//*
//DSNUPROC.SORTWK01  DDUNIT=SYSDA,SPACE=(CYL,(5,2))
//DSNUPROC.SORTWK02  DDUNIT=SYSDA,SAPCE=(CYL,(5,2))
//DSUPROC.SORTOUT DD DSN=CAT.SORTOUT,DISP=(MOD,CATLG,CARTLG),
//           UNIT=SYSDA,SPACE=(CYL,(5,1))
//DSNUPROC.SYSUT1 DD DSN=CAGT.SYSUT1,DISP=(MOD,DELETE,CATLG),
//           UNIT=SYSDA,SPACE=(CYL,(5,1))DCB=BUFNO=20
//DSNUPROC.SYSDISC DD DSN=CAT.SYSDISC,DSIP=(MOD,DELETE,CATLG),
//           UNIT=SYSDA,SPACE=(CYL,(2,1))
//DNSPROC.SYSERR DD DSN=CAT.SYSERR,DSIP=(MOD,DELETE,CATLG),
//           UNIT=SYSDA,SAPCE=(CYL,(2,1))
//DSNUPROC.SYSREC00 DD DSN=CAT.LOAD.INPUT.DATASETA,DSIP=SHR,
//                    DCB=BUFNO=20
//DSNUPROC.UTPRINT DD SYSOUT=X
//*
//* INPUT UTILITY STATEMENTS
//*
//DSNUPROC.SYSIN    DD *
   LOAD DATA REPLACE INDDN SYSREC00 LOG NO
   INTO TABLE DBNAME1.XXXX
      (ACTNUM        POS (1)   SMALLINT,
       ACTKWDY       POS (3)   CHAR (6),
       ACTDESCRIP    POS (10)  VARCHAR
)

/*
//****************************************************************
```

The sort work data sets need to be assigned in the JCL only as sort work data sets and are not
Dynamically allocated. With V5 and up it is possible to create an inline copy while loading,
use the COPYDDN and RECOVERDDN key words.

5.0.6 DB2 OS390 Unload Utility

One option for creating a readable sequential unload data set for DB2 tables (without writing
an application program) is the DSNTIAUL program. The REORG utility and
the UNLOAD ONLY option also can be used to unload DB2 data from a tablespace.

DSNTIAUL is a DB2 application program written in assembler. It can unload the data
from one or more DB2 tables or views into a sequential data set. The LOAD utility can
then use this data set. Additionally, DSNTIAUL can produce the requisite control cards
for the LOAD utility to load the sequential dataset back to the specific table.

```
//DB2JOBU  JOB (UILITY),'UNLOAD', CLASS=X,MSGCLASS=X,
//               NOTIFY=USER
//*
//*   DB2 UNLOAD PROGRAM DSNTIAUL
//*
//JOBLIB DD  DSN=DSN.LOAD.LIBRARY,DISP=SHR
//UNLOAD  EXEC  PGM=IKJEFT01,DYNAMNBR=20,COND=(4,LT)
//SYSTSPRT DD SYSOUT=*
//SYSTSIN   DD *
DSN SYSTEM(DSN)
RUN PROGRAM(DSNTIAUL) PLAN(DSNTIAUL) -
LIB('DSN.RUN.LOADLIB')
/*
//SYSPRINT  DD  SYSOUT=*
//SYSUDUMP  DD SYSOUT=*
//SYSREC00  DD DSN=THE.UNLOAD.DATASET,DSIP=(,CATLG,DELETE),
//            UNIT=SYSDA,SPACE=(CYL,(1,1)),DCB=BUFNO=20
//SYSPUNCH  DD  DSN=DEPT.RELOAD.UTILITY,DISP=(,CATLG,DELETE),
//              UNIT=SYSDA,SAPCE(TRK,(1,1)RLSE)
//SYSIN       DD *
DBNAME1.TABLE
/*
//
```

AFTER RUNNING:- You will get "PUNCHED" OUTPUT:-

```
LOAD DATA INDDN SYSREC00 LOG NO INTO TABLE
      DBNAME1.TABLE
(
   COL1  POS ( 1 )
```
Followed by all the "cards" for each column in the table.
In the SYSPUNCH Dataset.
```
)
```

5.0.7 DB2 OS/390 Merge Utility

```
//DB2JOB1  JOB (UILITY),'FULL IMAGE COPY', CLASS=X,MSGCLASS=X,
//                NOTIFY=USER
//*
//*   DB2 MERGE COPY
//*
//COPY  EXEC DSNUPROC,SYSTEM=DSN,UID='USERID',UTPROC="
//*
//DSNUPROC.SYSUT1 DD DSN=CAT.SYSUT1,DISP=(MOD,CATLG,CATLG),
//            UNIT=SYSDA,SPACE=(CYL,(10,1)),DCB=BUFNO=20
//DSNUPROC.SYSCOPY1 DD DSN=CAT.FULLCOPY.SEQ.DATASET1(+1),
//          DISP=(MOD,CATLG),DCB=SYS1.MODEL,
//          SPACE=(CYL,(5,2),RLSE),UNIT=3390
//DSNUPROC.SYSCOPY2 DD DSN=CAT.FULLCOPY.SEQ.DATASET2(+1),
//          DISP=(MOD,CATLG),DCB=SYS1.MODEL,
//          SPACE=(CYL,(5,2)RLSE),UNIT=3390
//* THE FIRST MERGE COPY CREATES A NEW FULL IMAGE COPY
//* THE SECOND MERGE COPY CREATES A NEW INCREMENTAL COPY
//* FOR THE NAMED TABLESPACE
//DSNUPROC.SYSIN        DD    *
    MERGECOPY  TABLESPACE DBNAME1.TSNAME1
              DSNUM ALL NEWCOPY YES
              COPYDDN SYSCOPY1
    MERGECOPY TABLESPACE DBNAME1.TSNAME2
              DSNUM ALL NEWCOPY NO
              COPYDDN SYSCOPY2

//*
//****************************************************************
```

Merge Copy Guidelines

1. Merge incremental copies asap.
2. Use Mergecopy to create Full Image Copies.
3. Specify the SYSUT1 Data Set - to avoid rerunning Mergecopy. Incase Mergecopy is unable to allocate all the datasets needed for the merge.
4. Mergecopy produces a message indicating the number of existing datasets and the number of Merged data sets. If the numbers are not equal run Mergecopy again.

6.0 DB2 Display

Use the DISPLAY STATISTICS command to obtain statistics for DB2 transactions. The information provided by this command is an accumulation of statistical counters if the CICS attachment to DB2 is activated with the DSNC STRT command. Directly after DB2 is attached to CICS all these numbers are zero.

The following counts are kept:-

TRAN	Transaction name associated with this RCT entry.
PLAN	Plan name associated with this RCT entry.
CALLS	SQL executions issued by transactions.
COMMITS	Number of COMMITS by transactions associated with the RCT entry.
ABORTS	Number of aborts including both ABENDS and ROLLBACKS.
AUTHS	Number of sign-ons for transactions associated with the RCT entry.
W/P	Number of times any transaction had to wait for an available thread.
HIGH	Highwater mark for the number of threads needed by any transaction.

7.0 Data Warehouse - Overview

A data warehouse is best defined by the type of data stored in it and the people who use that data. A data warehouse is designed for decision support providing easy access to relevant data and reducing data contention. It is separated from day-to-day OLTP applications that drive the core business rather than by computer processes. The data warehouse classifies information by subjects of interest to business analysts, such as customers, products and accounts. Data in the warehouse is seldom updated instead it is inserted (or loaded) and then read multiple times.

Warehouse information is historical in nature, spanning transaction over many months and years. For this reason, warehouse data is usually summarized to make it easier to scan and access. Redundant data can be included in the data warehouse to present the data in an easily understood grouping.

Fact Tables and dimension tables make for a condensed summary of information that may be accessed by queries
and printed by any number of tools "down stream".

7.0.1 A look at how the Data Warehouse process works

Enterprise source data feeds which are usually Flat or VSAM files are passed through an ETL (Extract, Transform and Load) process which creates other flat or VSAM files or directly populate Staging Tables. These Staging Tables usually contain selected but unprocessed data. The Staging Tables are then accessed by the Applications or tools like EVOKE and/or Data Stage and processed, transformed and expanded into other tables (including Dimension and Fact (Summary Tables)) in other databases, known as the Transaction Data Store(s) (TDS) or ODS (Operational Data Store). The TDS or ODS are in essence the Data Warehouse. The TDS/ODS supply data for various "down stream" reporting processes and invariably these reporting sub-systems are executed by either PC or Mainframe processes (e.g.:- Brio, Essbase, Cognos, Mainframe Reporting Programs etc.). Queries run against the Dimension and Fact Tables usually Join these tables to produce a result set.

7.0.2 Data Warehouse Summary

A data warehouse is a collection of data that is:-
Usually separate from normal operational systems.
Accessible / available for queries or reports (sometimes through the use of third party vendor products).
Subject oriented by business function (a.k.a. Subject Matter).

Integrated, named and defined. Sometimes associated with specific periods of time. Sometimes designated as static, or non-volatile, such that updates are not made.

There are no hard and fast rules for data warehouses, no fixed rule of thumb. As we move forward in learning how to best use these functions, it becomes apparent that the key to all of this is flexibility. Different shops have different ways of achieving the same goal, however, one must always bear in mind the performance objectives and service level agreements that each shop/group has and this dictates what will govern the structure and access parameters that are critical to success.

8.0 OS/390 Performance Monitors

8.0.1 Explain

Use Explain to see the access paths that DB2 uses when executing your SQL query. These access determinations are made by the DB2 Optimizer. These access path determinations make up what is known as the Plan Table.

8.0.2 Plan Table Columns.

Tname - Name of table being accessed.

Tabno - IBM use Only.

Accesstype - Indicates the method of accessing the table.

I	-	Indexed Access
I1	-	One-fetch Index scan
R	-	Tablespace scan
N	-	Index access with an IN predicate
M	-	Multiple index scan
MX	-	Specification of the index name for multiple index scan
MI	-	Multiple index access by RID intersection
MU	-	Multiple index access By RID Union
Blank	-	Row applies to QBLOCKNO 1 of an INSERT or DELETE statement using a cursor with the WHERE CURRENT OF clause.

Matchcols	Contains integer value with the number of index columns used in an Index scan when ACCESSTYPE is I, I1, N or MX. Otherwise 0.
Accesscreator	Indicates the creator of the index when Accesstype is I, I1, N or MX
Accessname	Name of index used when Accesstype is I, I1, N or MX
Indexonly Y	indicates that access to the index is sufficient to satisfy the query. N indicates that access to the tablespace is also required.
Sortn_uniq Y	Indicates that a sort must be performed on the new table to remove duplicates.
Sortn_join Y	indicates a sort must be performed to accomplish a merge scan join.

Sortn_Orderby Y indicates a sort must be performed to order
                 rows.
TSLOCKMODE   Contains the lock level applied to the new,
                 it's tablespace, or partitions. If the isolation
                 level can be determined at BIND time the values
                 can be as follows:-
                      IS   Intent share lock
                      IX   Intent exclusive lock
                      S   Share lock
                      U   Update lock
                      X   Exclusive lock
                      SIX Share with intent exclusive lock
                      N   No lock (UR isolation level)

8.0.3 DB2 PM

DB2-PM   - Most widely used, produces many reports, and
traces.

IBM's DB2-PM is the most widely used batch performance monitor
for DB2, also, DB2-PM provides an Online component.

DB2 PM can generate many categories of performance reports,
known as report sets. A brief description of each report set
follows:-

Accounting - Summarizes the utilization of DB2 resources such
as CPU and Elapsed time, SQL use, buffer use and locking.

Audit - Tracks the access of DB2 resources. Provides information on
Authorization failures, GRANTs and REVOKEs, access to Auditable
tables, SET SQLID executions, and utility execution.

I/O Activity - Summarizes DB2 reads and writes to the bufferpool,
EDM pool, active and archive logs, and the BSDS.

Record Trace - Displays DB2 trace records from the input
source.

SQL Trace - Reports on each activity associated with each SQL
Statement.

Statistics - Summarizes the statistics for an entire DB2
subsystem.

Summary - Reports on the activity performed by DB2-PM to produce the Requested reports.

System Parameters - Creates a report detailing the values assigned by DSNZPARMS.

With Long Accounting Trace, you will get the:- Start Time, End Time, Elapsed Time and CPU Time.

You can start a trace from the Display panel or by requesting Operations to do it.

Requesting Operations example of E-Mail to Operations:-

"We would like classes 1,,2,3,5,7,8 and 9 to be active. The destination should be SMF and the Plan should be DSNESPCS which is the SPUFI Plan."

Start a Trace using the DB2I Display Panel:-

START TRACE(ACCTG) CLASS(1,2,3,5,7,8,9) PLAN(DSNESPCS) DEST(SMF)

8.0.4 DB2 PM Possible Problem Areas

All performance problems are caused by some kind of change:-

Physical Changes to the environment such as a new CPU, new DASD devices, or different tape drives.

Installing a new version or release of the Operating System.

Changes to system software, such as a new release of a product (for example QMF, CICS regions or an IMS SYSGEN) or a completely new product e.g. DFHSM or even a new release of DB2 which can result in access paths and the utilization of new DB2 features.

Changes to the DB2 engine from maintenance releases which can change the optimizer.

Changes to system capacity. More or fewer jobs can be executing concurrently when the performance problem occurs.

Environment changes, such as the implementation of client/ server programs or the adoption of data sharing.

Database changes. This involves any DB2 object, ranging from adding a new column to dropping and re-creating an object.

Changes to the application development. Usage of check constraints or use of stored procedures.

Changes in the application code.

8.0.5 DB2 - Trace Records

Locking Report Set

EXPLAIN

RUNSTATS

I/O Activity Report

Audit Report Set

Bufferpool Information

SQL Trace

9.0   DNZPARMS

When DB2 is installed, the DSNZPARMs are set and implemented. Sometimes known as the ZPARMs, these parameter settings cover numerous limits and values for a host of performance related DB2 activities. The names and meanings of a few of these parameters are listed below:-

IDFORE – Max users in TSO connected to DB2

IDBACK – Controls number of background jobs accessing DB2

MAXDBAT – Maximum number of concurrent distributed threads that can be active at one time.

EDMPOOL – All package table information is loaded into this portion of memory called the EDMPOOL (Environmental Descriptor Management Pool). This parameter specifies the size of the EDM Pool.

NUMLKTS – Specifies the threshold for the number of page locks that can be held at the same time for one table space by any single application thread.

NUMLKUS – Specifies the threshold for the total number of page locks across all table spaces that can be held for a single DB2 application. If the NUMLKUS is exceeded by an application trying to acquire a lock, a "resource not available" (-904) message is received by the application.

10.0   OS/390 Utilities

10.0.1   OS/390 Quiesce Utility

The QUIESCE Utility is used to record a point of consistency for a tablespace, partition, tablespace set, or list of tablespaces and tablespace sets. QUIESCE ensures that all tablespaces within the scope of the QUIESCE are referentially intact. It does this by externalizing all data modifications to DASD and recording log RBAs (Relative Byte Address) and LRSNs (Log Record Sequence Number) in the SYSIBM.SYSCOPY DB2 Catalog table, indicating a point of consistency for future recovery. This is called the QUIESCE POINT. Running QUIESCE improves the probability of a successful RECOVER or COPY."

<u>Sample JCL</u>

```
//DB2JOB1  JOB (UILITY),'QUIESCE', CLASS=X,MSGCLASS=X,
//               NOTIFY=USER
//*
//*   DB2 QUIESCE UTILITY
//*
//*  STEP 1:- STARTUT – Start all tablespaces in the tablespace set in utility only mode
//*  STEP 2:- QUIESCE  - Quiesce all the tablespaces in the tablespace set
//*  STEP 3:- STARTRW – Start all tablespaces in the tablespace set in read/write mode
//*
//STARTUT  EXEC PGM=IKJEFT01,DYNAMNBR=20
//STEPLIB     DD  DSN=LOAD.LIBRARY.NAME,DISP=SHR
//SYPRINT     DD  SYSOUT=*
//SYSTSPRT  DD  SYSOUT=*
//SYSOUT      DD  SYSOUT=*
//SYSUDUMP DD  SYSOUT=*
//SYSTSIN     DD  *
DSN SYSTEM (DSN)
-START DATABASE (DBNAME1) ACCESS (UT)
END
/*
//QUIESCE EXEC  DSNUPROC,SYSTEM=DSN,UID'USERID',UTPORC="
//               COND=(0,NE,STARTUT)
//DSNUPROC.SYSIN DD  *
                QUIESCE TABLESPACE DBNAME1.TSNAME1
                        TABLESPACE DBNAME1.TSNAME2
                        TABLESPACE DBNAME1.TSNAME3
                        TABLESPACE DBNAME1.TSNAME4
                        TABLESPACE DBNAME1.TSNAME5
```

```
/*
//STARTRW EXEC PGM=IKJEFT01,DYNAMNBR=20,COND=EVEN
//STEPLIB DD DSN=LOAD.LIBRARY.NAME,DISP=SHR
//*
//SYSPRINT  DD  SYSOUT=*
//SYSOUT    DD  SYSOUT=*
//SYSTSIN   DD  *
DSN SYSTEM (DSN)
-START DATABASE (DBNAME1) ACCESS (RW)
END
```

10.0.2 DB2 OS/390 Recover Utility

RECOVER can be used to recover tablespaces or indexes by
restoring data from an image copy data set and then applying
subsequent changes from the log files.

The RECOVER TABLESPACE utility restores tablespaces to a current
or previous state. It first reads the DB2 Catalog to determine
the availability of full and incremental image
copies, and then reads the DB2 Logs to determine interim
modifications. The utility then applies the image copies and the
log modifications to the tablespace dataset being recovered.
The DBMS maintains the recovery information in the DB2 Catalog.
This enables the RECOVER utility to automate tasks such as:-

1. Retrieving appropriate image copy names and
   volume serial numbers
2. Retrieving appropriate log data set names and
   volume serial numbers
3. Coding the DD Statements for each of these in
   the recover JCL

Data can be recovered from a single page, pages that contain
I/O errors, a single partition of a partitioned tablespace, or
a complete tablespace.

Recover to a previous point can be accomplished by specifying
a full image copy or a specific log RBA. Recovery to the
current point can be accomplished by simply specifying only the
Tablespace Name as a parameter in the RECOVER utility.

Example of Recover JCL

```
//DB2JOB1  JOB (UILITY),'RECOVER', CLASS=X,MSGCLASS=X,
//               NOTIFY=USER
//*
//*   DB2 RECOVER UTILITY
//*
//* 1. RECOVER TABLESPACE
//* 2. RECOVER INDEX
//*
//RCVR EXEC  DSNUPROC,SYSTEM=DSN,UID'USERID',UTPORC="
//DSNUPROC.SYSIN  DD *
      RECOVER TABLESPACE DBNAME1.TSNAME1 DSNUM ALL
      REBUILD INDEX(ALL) TABESPACE DBNAME1.TBNAME1


/*

To recover to Image copy dataset:- RECOVER TABLESPACE DBNAME1.
TSNAME1

TOCOPY CAT.FULLCOPY.DATASET

To recover to the Log RBA:- RECOVER TABLESPACE DBNAME1.TSNAME1
TORBA X'0000EF2C66F4'
/*
//
```

Sample SQL to Identify Tape DSN on to which a table space was backed up

```
SELECT DSNAME, TSNAME, ICDATE, ICTIME FROM SYSIBM.SYSCOPY
WHERE DBNAME = 'DBNAME' AND TSNAME = 'TSNAME'
AND TIMESTAMP = (SELECT MAX (TIMESTAMP) FROM SYSIBM.SYSCOPY
WHERE DBNAME = 'DBNAME' AND TSNAME = 'TSNAME' AND ICTYPE = 'F')
```

10.0.3 DB2 OS/390 REORG Utility

```
The REORG utility can be used to reorganize DB2 tablespaces and
indexes, thereby improving the efficiency of access to those
objects. Reorganization is required periodically to ensure
that the data is situated in an optimal fashion for subsequent
access. Reorganization reclusters data, resets free space to
the amount specified in the CREATE DDL and deletes and redefines
the underlying VSAM datasets for STOGROUP defined
Objects.
```

Example of REORG JCL

```
//DB2JOB1  JOB (UILITY),'REORG', CLASS=X,MSGCLASS=X,
//                      NOTIFY=USER
//*
//*     DB2 REORG UTILITY
//*
//*
//UTIL EXEC  DSNUPROC,SYSTEM=DSN,UID'USERID',UTPORC=''
//*
//* WORK DATASETS
//*
//DSNUPROC.SORTWK01   DDUNIT=SYSDA,SPACE=(CYL,(5,1))
//DSNUPROC.SORTWK02   DDUNIT=SYSDA,SAPCE=(CYL,(5,1))
//DSUPROC.SORTOUT DD DSN=CAT.SORTOUT,DISP=(MOD,CATLG,CARTLG),
//                  UNIT=SYSDA,SPACE=(CYL,(2,1))
//DSNUPROC.SYSUT1 DD
//                DISP=CAGT.SYSUT1,DISP=(MOD,DELETE,CATLG),
//                UNIT=SYSDA,SPACE=(CYL,(2,1))DCB=BUFNO=20
//DSNUPROC.SYSREC DD DSN=INPUT.DATASETD,DSIP=SHR,
//                  DCB=BUFNO=20
//DSNUPROC.UTPRINT DD SYSOUT=X
//*
//* INPUT REORG UTILITY STATEMENTS
//*
//DSNUPROC.SYSIN    DD  *
       REORG TABLESPACE DBNAME1.TSNAME1

/*
//**************************************************************
```

10.0.4 DB2 OS/390 RECOVER UTILITY

The Recover Utility can recover table spaces or indexes by
restoring data from a previous image copy and then (from that
point in time) applying changes recorded on the log files.

An example of JCL for a Full Recovery follows:-

```
//DB2JOB1   JOB (RECUTIL), 'RECOVER FULL', CLASS=K,
//          MSGCLASS=X,NOTIFY=XXXXXXXX
//*
//*   DB2 FULL RECOVERY
//*
//RECOV   EXEC  DSNUPROC,SYSTEM=DSN,UID='FULLR',UTPROC=''
```

```
//*
//*                        1. RECOVER A TABLESPACE
//*                        2. RECOVER THE INDEXES IN THE TABLESPACE
//*
//DSNUPROC.SYSIN     DD  *
     RECOVER TABLESPACE DATABAS1.TBLESPAC1 DSNUM ALL
     REBUILD INDEX(ALL) TABLESPACE DATABAS1.TBLESPAC1
//*
//
```

An example of JCL for a Partial Recovery follows:-

```
//DB2JOB2 JOB (RECUTIL2), 'RECOVER PART', CLASS=K,MSGCLASS=X,
//          NOTIFY=XXXXXXXX
//*
//* DB2 PARTIAL RECOVERY
//*
//RECPAR EXEC DSNUPROC,SYSTEM=DSN,UID='PRTREC',UTPROC=''
//*
//* 1, RECOVER TABLESPACE TO IMAGE COPY  2. RECOVER TO LOG RBA
//*
//DSNUPROC.SYSIN   DD *
   RECOVER TABLESPACE DATABAS1.TBLESPAC1 TOCOPY
   CAT.FULLCOPY.DATASET.G0002V00
   RECOVER TABLESPACE DATABAS1.TBLESPAC2 TORBA X'0000DE3D55E2'
//**
```

An example of JCL for a Index Space Recovery follows:-

This "recover index space" function is the same as the "recover table space" function except that it operates on DB2 indexes and not DB2 table spaces. Remember, as of Version 6.1 REBUILD INDEX is used to rebuild the indexes from the table data so this utility is now less used.

```
//DB2JOBR JOB (RECUTIL), 'DB2 RECVR INDEX',MSGCLASS=M,CLASS=K,
//          NOTIFY=XXXXXXXX
//*
//*         RECOVER INDEX UTILITY
//*
//RECUTIL    EXEC DSNUPROC,SYSTEM=DSN,UID='RECVRIDX',UTPROC=''
//*
//DSNUPROC.SYSIN   DD *
   RECOVER INDEXSPACE DATABAS1.XISPAC1
```

```
//*
//
```

<u>10.0.5 OS/390 Rebuild Index</u>

Note:- The REBUILD INDEX utility can be used to re-make indexes from the current data in the table. When an Index is defined as COPY NO it is always recovered from the actual table data. When an index is defined as COPY YES, it can be recovered from an image copy or rebuilt from the table data. The REBUILD INDEX function scans the table and regenerates the index based on current data.

```
//DB2JOBRI    JOB (UTILRBIX),'REBUILD IDX)',MSGCLASS=X,CLASS=K,
//              NOTIFY=USER
//*
//*          REBUILD INDEX
//*
//UTILRIDX    EXEC  DSNUPROC,SYSTEM=DSN,UID='RBLDIDX',UTPROC="
//*
//DNSUPROC.SORTWK01  DDUNIT=SYSDA,SPACE=(CYL,(5,2))
//DNSUPROC.SORTWK02  DDUNIT=SYSDA,SPACE=(CYL,(5,2))
//DNSUPROC.SYSUT1  DD  DSN=&&SYSUT1,
//              UNIT=SYSDA,SPACE=(CYL,(5,2)),DCB=BUFNO=30
//DSNUPROC.UTPRINT  DD  SYSOUT=X
//*
//DSNUPROC.SYSIN      DD *
  REBUILD INDEX (DATABAS1.XIDX01)
  REBUILD INDEX (DATABAS1.XIDX02)  DSNUM 5   <---- PARTITION 5 ONLY
  REBUILD INDEX (ALL) TABLESPACE DATABAS1.TBLESPA1
//*
//
```

The MODIFY RECOVERY utility gets rid of recovery information from two system catalog tables, i.e the SYSCOPY and SYSLGRNX tables.

You can remove recovery information either by deleting rows that are older than a specified number of days or before a specified date.

<u>10.0.6   MODIFY RECOVERY Utility</u>

```
//DB2JOBMR    JOB (MRUTIL),'MODRCVR',MSGCLASS=X,CLASS=K,
//            NOTIFY=XXXXXXXX
//*
//*           MODIFY RECOVERY
//*
//MRUTIL      EXEC  DSNUPROC,SYSTEM=DSN,UID='MODREC',UTPORC="
//*
//*           1. DELETE SYSCOPY DATA GREATER THAN 45 DAYS OLD
//*           2. DELETE SYSCOPY DATA BEFORE JANUARY 1, 1990
//DSNUPROC SYSIN  DD  *
   MODIFY RECOVERY TABLESPACE DATABAS1.TBLESPC1 AGE 45
   MODIFY RECOVERY TABLESPACE DATABAS1.TBLESPC2 DATE (19900101)
//*
//
```

<u>10.0.7   RUNSTATS Utility</u>

After every REORG, you should run the RUNSTATS utility. The RUNSTATS utility keeps the Catalog Tables "in synch" with the current state of the tables.

You may use the RUNSTATS utility to:-

1. Update the DB2 Catalog with all the statistics that have been gathered.
2. Update the DB2 Catalog with DBA monitoring statistics.
3. Update the DB2 Catalog with only DB2 Catalog.
4. Produce a Statistics Report without updating the DB2 Catalog tables.

There are three forms of RUNSTATS.

a. Operates at the table space level.
b. Operates at the table and column level.
c. Operates at the index level.

<u>Example RUNSTATS JCL follows:-</u>

```
//DB2RUNST    JOB (UTILRUNS),'RUNSTATS',MSGCLASS=X,CLASS=K,
//            NOTIFY=XXXXXXXX
//*
```

```
//*                  RUNSTATS UTILITY
//*
//*           1. ACCUMULATE STATISTICS FOR THE TABLE SPACE
//*           2. ACCUMULATE STATISTICS FOR TABLE AND COLUMNS
//*           3. ACCUMULATE STATISTICS FOR A SPECIFIC INDEX
//*
//UTILRUN  EXEC DSNUPROC,SYSTEM=DSN,UID='RUNSTATS',UTPROC=''
//*
//DSNUPROC.SYSIN  DD  *
      RUNSTATS  TABLESPACE DATABAS1.TBLESPC1
         INDEX (ALL)  SHRLEVEL REFERENCE
      RUNSTATS  TABLESPACE DATABAS1.TBLESPC2
         TABLE (TABLE1.EMPLOYEE)
         COLUMN (LASTNM,MIDINIT,FIRSTNM,EDUC,SALARY)
         SHRLEVEL REFERENCE
      RUNSTATS INDEX (DATABAS1.XIDX01)
/*
//
```

```
Tips for running RUNSTATS:-

   1. Run during off peak hours.
   2. Limit columns to 10 or less.
   3. Always run RUNSTATS after table or data changes.
   4. Use RUNSTATS to produce Statistics Reports.
```

10.0.8 DB2 Mainframe – Assorted Utilities

```
DSNJU003 - Change Log Inventory Control - modifies the bootstrap
data set (BSDS). It's primary function is to add or delete
archive logs for the DB2 Subsystem. e.g. PGM=DSNJU003.
NEWLOG DSNAME=XXX.YYYY
NEWLOG DSNAME=XXX.YYY2


DSNJU004 - The Print Log Map utility, is used to display the
status of the logs in the BSDS. e.g. PGM=DSNJU004.  SYSUT1 DD
DSN=DB2CAT.BSDS01
SYSPRINT DD SYSOUT=*


DSNJLOGF - Is the DB2 Log Pre-format Utility. It pre-formats
DB2 active log data sets. DSN1JLOGF  SYSUT1 DD DSN=DSNXXX.
LOGCOPY1, PGM=DSN1JLOGF  SYSUT1 DD DSN=DSNXXX.LOGCOPY2
```

DSN1CHKR - The Catalog Integrity Verification utility, verifies the integrity of the DB2 Catalog and DB2 Directory. PGM=DSN1C HKR,PARM='FORMAT'
SYSUT1 DD DSN=DB2CAT.XXX.YYY,Note:- The SYSUTILX and SYSLGRNX tables are not checkable using DSN1CHKR. This is true even though the predecessors to these tables were checkable (SYSUTIL prior to DB2 V3 and SYSLGRNG prior to DB2 V4). Execute DSN1CHKR weekly to catch problems early, before they affect program development and testing in your test DB2 subsystems or business availability and production processing in your production DB2 environment.
Also run DSN1COPY with the check option against all DB2 Catalog indexes and table spaces. Run the Check Index utility against all catalog indexes.

DSN1COMP - This is the compression analyzer service aid. It can be used to approximate the results of DB2 data compression. DSN1COMP can be run on a table space dataset, a sequential dataset containing a DB2 table space or partition, a full image copy data set, or an incremental image copy data set. It provides compression statistics, % bytes saved,total pages required, pages saved and average size of compressed row.
PGM=DSN1COMP,PARM='ROWLIMIT(20000)'SYSUT1   DD   DSN=DB2CAT. DSNDBC.XXXX,DISP=OLD etc, numerous parameters can be supplied e.g. FREEPAGE, PCTFREE, FULLCOPY , INCRCOPY, REORG, ROWLIMIT

DSN1COPY - This utility is the Offline Table space Copy Service, and has a multitude of uses. For example, it can be used to copy datasets or check the validity of Table space and index pages. Another use is to translate DB2 Object Identifiers for the migration of objects between DB2 subsystems or to recover data from accidentally dropped objects. DSN1COPY can also print Hexadecimal Dumps.

Here are all the functions of DSN1COPY:-

Create a sequential data set copy of a DB2 table space or index data set.
Create a sequential data set copy of another sequential data set copy produced by DSN1COPY.
Create a sequential data set copy of an image copy data set produced using the DB2 Copy utility, except for segmented table spaces (The DB2 Copy utility skips empty pages, thereby rendering the image copy data set incompatible with DSN1COPY).
Restore a DB2 table space or index using a sequential data

set produced by DSN1COPY.
Restore a DB2 table space using a full image copy data set produced using the DB2 Copy utility.
Move DB2 data sets from one disk pack to another to replace DASD such as migrating from 3380s to 3390s).
Move a DB2 table space or index space from a smaller data set to a larger data set to eliminate extents, or move a DB2 table space or index space from a larger data set to a smaller data set to eliminate wasted space.
DSN1COPY runs as a batch job, so it can run as an Offline utility when the DB2 subsystem is inactive. If it runs when the DB2 subsystem is active then the objects it operates on should be stopped to ensure DSN1COPY creates valid output. DSN1COPY does not communicate with DB2.
e.g. Copy one dataset to another using DSN1COPY:-

```
-STOP DATABASE (DSNXXXXX) SPACENAM(DSNYYYYY)
EXEC PGM=DSN1COPY,PARM='CHECK'
SYSUT1 DD DSN=DB2CAT.XXXXX,DISP=OLD
SYSUT2 DD DSN=OUTPUT.SEQ.DATASET,DIS=OLD
-START DATABASE (DSNXXXXX) SPACENAM(DSNYYYYY)
```

One of the best features of DSN1COPY is it's ability to modify the internal Object Identifier (OBID) stored in the DB2 table space and index data sets as well as in data sets produced by DSN1COPY and the DB2 COPY utility.
When you specify OBIDXLAT option, DSN1COPY reads a data set specified by the SYSXLAT DD Statement. This data set lists the source and target DBIDs, PSIDs or ISOBIDs and OBIDs. The DNS1COPY utility can only translate up to 500 record OBIDs at a time.

Do not use DSN1COPY on Log Data Sets - certain options can invalidate the Log data.

Use the DSSIZE parameter to specify data sets that exceed 2GB. The DSSIZE parameter must exactly match the size of the table space defined.

DSN1SDMP - Is the IFC selective dump utility. This is actually an application program. It can force system dumps. It must be run under TSO program IKJEFT01.

11.0 DB2 OS/390 Frequently returned SQL CODES

Code	Description
-805	DBRM or PACKAGE NAME (location name, collection-id.dbrmname.consistency) token not found in PLAN.
-807	Access denied. Package isn't enabled for access from (connection-type connection-name).
-808	The CONNECT semantics that apply to an application process are determined by the first CONNECT statement executed (successfully or unsuccessfully) by the application process.
-811	Execution of an embedded SELECT statement has resulted in a result statement containing more than one row. Alternatively, a subquery contained in a basic predicate has produced more than one value.
-817	The SQL statement can't be executed because the statement will result in a prohibited Update operation.
-818	The precompiler generated time stamp in the load module is different from the bind timestamp built from the DBRM.
-819	The SYSIBM.SYSVTREE.VTREE is a varying length string column that contains the parse trees of views. In processing a view, the length control field Of it's parse tree was found to be zero.
-820	The SQL statement can't be processed because (catalog table) contains a value That is not valid in this release.
-904	Unavailable Resource.
-911	The current unit of work has been rolled back due to a deadlock or time-out.
-913	Unsuccessful execution caused by deadlock or timeout.

12.0 DB2 OS/390 Stored Procedures

```
CREATE PROCEDURE proc-name
IN/OUT/INOUT                parm-name data type
SPECIFIC                          specific name
RESULTS SETS             0
or RESULTS SETS          integer
EXTERNAL NAME          implementation
LANGUAGE                    C / Java
NULL CALL
PARAMETER STYLE       DB2DARI or DB2GENERAL
NOT DETERMINISTIC
or DETERMINISTIC
STAY RESIDENT YES
PROGRAM TYPE SUB
FENCED
or NOT FENCED
```

The **CREATE** function stores the description of the procedure in the System Catalogs – Procedure or Routines and PROCPARMS.

EXTERNAL tells UDB how to find the executable program that implements the stored procedure. e.g. EXTERNAL NAME '/bank/ bin/stprocs!server1'

LANGUAGE JAVA – if your procedure is written in JAVA you must specify LANGUAGE JAVA and PARAMETER STYLE DB2GENERAL.

Otherwise LANGUAGE C PARAMETER STYLE DB2DARI

DETERMINISTIC - If the result of your Stored Procedure is uniquely determined by it's input parameters you can make this fact known to the system by specifying DETERMINISTIC.

STAY RESIDENT YES – The Stored Procedure stays in memory during the execution of the job, thus avoiding re-loads.

PROGRAM TYPE SUB - The Stored procedure is treated as a sub-routine, this is faster than specifying MAIN.

NULL CALL - If the Stored Procedure is invoked even when the input parameters are null, you can acknowledge this by specifying NULL CALL.

NOT FENCED - If your Stored Procedure is trusted and you wish
to run it in the same address space as the Database Manager you
may specify NOT FENCED. FENCED is safer and besides with FENCED
you can call REXX or CLI functions which you can't do with NOT
FENCED. Also, in order to create a NOT FENCED Stored Procedure
you must hold SYSADM or DBADM authority.

e.g.

```
                    CREATE PROCEDURE server1
                    (INOUT n_updates Integer,
                    EXTERNAL NAME 'stprocs!server1'
                    LANGUAGE C
                    DETERMINISTIC
                    PARAMETER STYLE DB2DARI;
```

Note:- The use of Global Temporary Tables (GTT) are common
place with the use of Stored Procedures, but for performance
reasons we found that we had to convert second and lower level
Stored Procedure programs to COBOL Programs. If this is the
case you may want to think about your GTTs. The instance of a
GTT exists until the remote server connection under which the
instance was created terminates or the Unit of work under which
the instance was created completes. When you execute a ROLLBACK
Statement, DB2 deletes the instance of the created temporary
table. This also happens when you do a COMMIT unless a Cursor
for accessing the created temporary table is defined WITH HOLD
and is open.

So if you are using COBOL Calls eliminate the use of GTTs if
possible or delete the data in the GTT after each run.

13.0   OS/390 Type 2 Indexes

Released with V6 in June of 1999.

This release also included large objects, triggers, user-defined functions, user defined data types and more. And, in V6, for the very first time IBM has removed features from DB2. As such your company must prepare it's DB2 sub-systems for V6 by removing the unsupported features from your installation.

1. Type 1 Indexes - being removed.

   Type 2 indexes are in do not use the Type 1 index sub
          pages.
   Type 2 indexes are supported for ASCII encoded tables.
   Type 2 can support row level locking
   Type 2 can support data sharing
   Type 2 can support Full Partition independence
   Type 2 can support uncommitted reads ISOLATION(UR)
   Type 2 can support UNIQUE WHERE NOT NULL
   Type 2 can support CPU and Sysplex parallelism

   To find all Type 1 indexes:- SELECT CREATOR, NAME
   FROM SYSIBM.SYSINDEXES WHERE INDEXTYPE = ' ';

2. Shared Read Only (SROD) with V2.3 is gone .

3. Recover Index feature of V5 replaced by REBUILD Index.

4. Host variables without colons - GONE!

5. Data set passwords using DSETPASS key word of CREATE
   TABLESPACE and CREATE INDEX disappears with V6.

   To find datasets with DSETPASS:-

```
            SELECT 'INDEX', CREATOR, NAME'
            FROM SYSIBM.SYSINDEXES
            WHERE DSETPASS <> '              '
            UNION ALL
            SELECT 'TSAPCE', DBNAME, NAME
            FROM SYSIBM.SYTABLESPACE
            WHERE DSETPASS <> '              '
```

6. STORED PROCEDURE REGISTRATION
   INSERT GONE, now CREATE STORED PROCEDURE

7. Feature APAR (Authorized Program Analysis Report) can
   install V6 features before actually installing V6.

14.0 OS/390 Set up a partitioned table space

Consider creating large partitioned table spaces which will support tables with more than 64 Giga Bytes of data, or more than 63 partitions. With large partitioned table spaces you can have up to 254 partitions of 4 Giga Bytes each , while normal (non-large) partitioned table spaces can have up to 64 partitions. This allows the size of partitioned tables to increase to roughly one terabyte. The maximum size of a data set for a non-partitioned Index on a LARGE partitioned table space is 4 Giga bytes. With a limit of 128 datasets, the maximum size of a nonpartitioned index is 512 Giga Bytes.

There are two parts to creating a partitioned table space on OS/390:-

1. Create the large partitioned table space:-

```
CREATE LARGE TABLESPACE EMPLY
IN DATABSE1
USING STOGROUP STGRP1
    PRIQTY 5000
    SECQTY 500
    ERASE NO

NUMPARTS 32
```

    .
    .
    . ←-   Define individual partitions 1 to 21 here
            (similar to Partition 22)

    .

```
(PART 22 USING STOGROUP STGRP1
    PRIQTY 3000
    SECQTY 300
    COMPRESS YES
```

    .
    . ←- Define individual partitions 23 to 32 here
    (similar to Partition 22)

    .

2.   Create a clustered index and specify key values for the partitions:-

```
      CREATE TYPE 2 INDEX EMPLYIDX
       ON EMPLY
      (EMPLEE ASC,  DEPT ASC)
       CLUSTER (PART 1 VALUES ('001',  '01' USING STOGROUP
           STGRP1,PART 2 VALUES ('099', '10') USING
              STOGROUP STGRP2, PART 3 VALUES ('198', '20')
              USING STOGROUP STGRP3,PART 4 VALUES ('296',
              '30') USING STOGROUP STGRP4,
                            .
                            .
                            .
```

define all the partitions in this way up to Partition 32 (in this case).

14.0.1 OS/390 Partition Load

Load Partition 22 of a 32 Partition Tablespace

```
LOAD DATA LOG NO INDDN SYREC00  ENFORCE CONSTRAINTS  ERRDDN
INTO TABLE XYZ.TABL001 PART 22 RESUME YES
(CON_LAST2_DIGITS POSITION (5:6) CHAR(2)
  .
  . ←---------- Specify other columns here as appropriate.
  .
  .)
```

15.0   OS/390 DB2 Tuning Tips

1.   Keep current with the technology. This is a fast paced Environment and new techniques and tools are becoming available at an ever increasing rate. Consequently, education is of the utmost importance. First, you have to understand what you are dealing with, this means the architecture of the version of DB2 you are using. Moreover you have to understand how to use features which are being used by and increasing number of users like Stored Procedures managed by the Work Load Manager, LOBs, CLOBs and BLOBs, Triggers, User Defined Functions and User Defined Ttpes.

2.   Have a plan or check list so you can find the problem in a systematic manner. Don't guess or "take a shot at the problem", it wastes every body's time including your own. Know how to use all the tools you have at your disposal, and use these tools to your benefit.

3.   If something does not make any sense, ask questions. Do not feel at all shy or embarrassed to ask questions. Asking questions could save you a lot of time and headaches.

4.   The DB2 Optimizer, brilliant as it is, does make mistakes! So be on the "look out" for any access paths that do not make sense.

5.   Check to see whether all the tables in an SQL query Have been REORGed recently, and RUNSTATS run. If not, then you must REORG and run RUNSTATS before testing the query.

6.   When matching tables, make sure that you use all the available keys. If you do not use all available primary key columns for a match, DB2 may do a table space scan which would take a long time. Your job is to influence DB2 to do an index scan on a match.

7.   Do not define redundant predicates. Define only what is needed, otherwise DB2 could take much longer to execute a WHERE clause than anticipated.

8.   The design of the Database must be good. Common sense tells you that, however, you will be very surprised to know that over 60% of Databases are not designed properly.

Look for simple bad design clues, like the primary key columns not being referenced in the sequence of the actual table columns, and the parent / child relationship in RI having key columns in an unmatched sequence.

9. The Database tables should be designed so that each table has a unique primary key (no duplicates). If duplicates "abound", DB2 may try and sort these tables, wasting a lot of time. Design your primary key so that no duplicates can exist.

10. Understand the SQL you are testing. It pays to study the SQL before you test using EXPLAIN or some other tool. If you really understand the SQL, this can save you much time.

11. Set up a comprehensive PLAN_TABLE and understand how to read it after an EXPLAIN. If you can accurately interpret the results in the PLAN Table you have made much progress. Look for unnecessary Table Space scans, sorts, duplicates and any other "thing" that does not look good or does not make sense. If you have no index scans this could have a negative effect on the performance of your query. Getting good at analyzing a PLAN Table only comes with experience, you do not have to be brilliant, just experienced!

12. I/O is the most detrimental feature to DB2.

13. Reducing I/O is the best scenario. Obviously, no I/O at all is the very best situation to be in, but these situations are rare. In addition to reducing I/O, you can do a few other things like ensuring you are using a large buffer pool. Check your buffer pool assignment, size and thresholds and make sure they are adequate. Only the DB2 Catalog and Directory should be in buffer pool BP0. In addition, your table spaces and indexes should be in separate buffer pools.

14. You must understand the nuances and intricacies of Locking. If you are involved in Performance Tuning this is a must. Make sure you do not use too much row level locking, because this will cause lock escalation. On the other hand, too little will risk long lock wait times.

15. For database tables with over a million rows or even smaller tables which have many concurrent users doing a lot of updates, make sure the table(s) are partitioned. This will always return better performance than non-partitioned tables. If there is a long running batch job using a table, partitioning this table can allow parallelism, which will translate into faster access times.

16. Be careful when creating indexes. Too many indexes can cause your query to run slower, not faster. Make sure the order of the columns defined in the index are in the same order as the table columns. Also, when a table is created there should not be any non key columns "splitting" the key columns, and no VARCHAR columns should be part of a key or index.

17. Make sure the PCTFREE and FREEPAGE parameters are used as appropriate. Always define your primary key as a clustered index. Make sure these parameters cater for the maximum size of variable records if you have these types of rows in a table.

18. Have you allowed enough space for fast growing tables?

19. Double check everything!!

20. Get friendly with the Developers, they usually turnout to be excellent sources of useful information. If the developers are using new technologies such as MQ Series, Java, XML or Websphere it will be useful for you to gain some background understanding of these technologies.

21. Check the performance across the network. Thus, it would benefit you to know someone in the Network Department to tell you about routers, gateways and bandwidth and to ascertain whether or not there are a sufficient number of these objects.

22. Check the statistics of the EDM Pool. Is your EDM Pool big enough?

23. Determine the number of users the application supports. If your system is using DB2CONNECT, then make sure the MAXBAT and CONDBAT parameters are set correctly and check the

DDF activity in the Performance Monitor Reports. Rebind programs with CURRENTDATA, ISOLATION and FOR FETCH ONLY.

24.   Check your Stored procedures are classified correctly.

25.   Check the WLM and DDF.

26.   You must also understand the application from a business point of view. If you do   not understand the business, you will be at a distinct disadvantage. Very often DB2 is not the culprit, it something outside DB2. But it is easy to blame DB2 because of it's "visibility". Check the DNZPARMS are correct for the use of DB2CONNECT. Are all those subsystems and objects outside DB2 free of blame for poor performance? Things like CICS, IMS, Token Ring, Gigabit, Ethernet, TCP/IP, SNA, Routers?

27.   Keep constantly monitoring performance! You should develop a Master Plan and monitor performance on a regular and systematic basis. Develop an EXCEL Spread Sheet to Log your results. Carefully monitor the results on your Spread Sheet and make sure Performance is not degrading even by a small bit. You could be on a slippery slope to a major slow down. If your efforts are constantly improving performance, let the "powers that be" know about it!

28.   Remember, Rome was not built in a day. Good DB2 Performance Tuner DBAs do not materialize from nowhere, either. It takes continual practice and training to become good at Performance Tuning. There is no genius involved, just lots of sweat and good experience. The "buck" stops with you, so you have to fix it and improve performance. Remember, don't blame anyone for any performance issues even though it may be blatantly obvious who the "culprit" really is, the motto of one great company springs to mind:- "<u>Fix the problem not the blame</u>".

16.0   OS/390 DB2 - What is the BSDS?

BSDS stands for **B**oot **S**trap **D**ata **S**et. This data set is a VSAM KSDS sequenced file that contains information which is vital to DB2, as follows:-

- A list of all active and archive log data sets used by DB2 in order that DB2 may track these log data sets. DB2 also uses the BSDS to find log records that satisfy log read requests during restart and recovery functions or even normal processing.

- For a log, the BSDS tells DB2 the RBA (relative byte address) range or the LRSN (log record sequence number) range in a data sharing environment, for every volume in which the log resides.

- The BSDS also tells DB2 which logs are full and which logs are available for re-use.

- The BSDS keeps track of all checkpoint activity, and uses this data for restart processing.

- In addition the BSDS contains the DB2 location name(s) (Distributed Data Facility communication record) and a table of IP addresses which are used to identify hosts within the TCP/IP network.

- The BSDS is critical to recovery in the event of system failure and because of it's importance, DB2 automatically creates two copies of the BSDS at installation time and places these copies on separate volumes.

17.0   The EDM Pool

What is the EDM Pool? The Environment Descriptor Management Pool is contains several parameters and tables specified during the installation process. The DSNTINST CLIST calculates the actual size of the EDM Pool based on the parameters specified on the DSNTIPD and DSNTIPE panels.

The EDM Pool contains:-

➢ Database Descriptors – DBDs

➢ Skeleton cursor tables – SKCTs

➢ Cursor Tables – CTs – copies of the SKCTs

➢ Skeleton package tables – SKPTs

➢ Package Tables – PTs or copies of the SKPTs

➢ An authorization block for each plan, excluding plans that were created speficying CACHESIZE(0)

➢ Skeletons of dynamic SQL if your installation has YES for the CACHE DYNAMIC SQL field.

18.0   The RID Pool

The Record Id. Pool is used for all record identifier processing. It is used for sorting RIDs during the following operations:-

List Pre-fetch, Single Index List Pre-fetch, Access via multiple indexes, Hybrid joins.

RID pool storage is also used when unique keys are enforced by DB2 while updating multiple rows.

Statements in SQL that use the RID Pool processing can reduce I/O resource consumption and elapsed time. Note, if there is not enough RID Pool storage, it is possible that the statement might revert to a table space scan.

To ascertain whether or not a transaction used the RID Pool, check the RID Pool Processing section of the DB2 PM accounting trace record.

The RID Pool, is shared by all concurrent work, is limited to approximately 1000 Mega Bytes. The RID Pool is created at system initialization time, however, no space is allocated until RID storage space is needed. RID space is allocated above the 16 Mega Byte line in 16 Kilo Byte blocks, until the maximum size specified during installation (panel DSNTIPC) is reached.

The formula for computing the RID Pool is:-

 # concurrent RID processing activities * average # of RIDS * 2 * 5 bytes per RID

19.0   OS/390 DB2 Migration

DB2 does not provide a feature to migrate DB2 objects from one subsystem to another. This can be accomplished manually by storing the CREATE DDL statements (and all subsequent ALTER statements) for future application in the other system. Manual processes such as this are error prone. Also, this process does not take into account the migration of table data, plans, DB2 security, packages, statistics and so on. DB2 object migration tools facilitate the quick migration of DB2 objects from one DB2 subsystem to another. They are similar to a table altering tool but have minimal altering capability (some interface directly with an alter tool or are integrated into a single tool). The migration procedure is usually driven by SPF panels that prompt the user for the objects to migrate.

Migration typically can be specified at any level. For example, if you request the migration of a specific database you also could migrate all dependent objects and security. Minimal renaming capability is provided so that database names, authorization ids., and other objects are renamed according to the standards of the receiving subsystem. When the parameters of the migration have been specified completely, the tool creates a job stream to implement the request DB2 objects in the requested DB2 subsystem.

A migration tool reduces the time required by database administrators to move DB2 databases from environment to environment (for example, from Test or QA to Production). Quicker turnaround results in a more rapid response to user needs, thereby increasing the efficiency of your business.

Typically, migration tools are the second DB2 tool that an organization requires (right after a DB2 Catalog Query Product). When evaluating migration tools, look for the following capabilities:- Should be able to run directly against the DB2 Catalog and/or against a copy of the DB2 Catalog.

Is able to run in the foreground (online) or background (batch).

Is able to migrate DB2 Objects, including plans and packages, triggers, stored procedures, and user defined functions, data and authorization

Is able to rename DB2 Objects during migration on the fly.

Is able to migrate all objects within a specific database easily
(for example, specify the database name and then automatically
migrate all dependent objects).

Is able to operate across all DB2 subsystems.

Is able to provide change control facility and the ability to
track versions of DB2 databases.

Is able to provide an object comparison facility to highlight
differences.

Allow you to restart a migration strategy at the point of
failure, or any other point as desired.

Is able to supply you with a complete record of what exactly
has been done.

20.0 OS/390 DB2 Performance Monitoring

1. First provide a comprehensive approach to monitoring.

   Monitor threads and address spaces by:-

   a. Batch reports run against DB2 trace records. While DB2 is running it actively traces information which can be used to monitor the performance of the DB2 Sub-system and also the applications being run.

   b. Online access to DB2 trace information and DB2 Control Blocks. This type of monitoring also can provide information on DB2 and it's subordinate applications.

   c. Sampling DB2 application programs as they run and analyzing which portions of the code use the most resources, is a very useful monitoring technique.

Remember do not go "hog wild" monitoring and tracing. These facilities use an absolutely huge amount of resources.

Plan and implement two types of monitoring:-

1. Ongoing performance monitoring.
2. Procedures for monitoring exceptions after they have been observed.

Do not try to hit a pin with a M1 tank!! Use the right tool for the job based on the type of problem you're monitoring.
This holds true for all DB2 problems. Be a "skilled surgeon" not a butcher.

Use Explain or the DB2 Catalog Reports or both.

Tuning should not consume your whole life, except of course if you happen to be a professional tuner! Establish your DB2 performance tuning goals in advance and stop when they have been achieved. Know when to pause and when to start tuning again.

John V. McLean

Types of performance monitoring

<u>DB2 trace Types</u>

Accounting DNZPARM or -START TRACE   Records performance
information about DB2 Application Programs.

Audit as above Provides information about DB2 DDL, security,
utilities and data modification.

Global as Provides information for the servicing of DB2

Monitor the above records data useful for Online.
Monitoring Performance as above Collect detailed data about
DB2 events. Enabling database performance analysis.
To pinpoint the cause of performance Problems.

Statistics   records   information   regarding   DB2   use   of
resources.

21.0 Helpful SQL Scripts

Remember to run STOSPACE and RUNSTATS before you run any of
these scripts.

```
-------------------------------------------------------------------
-- 1.  MONITOR THE SPACE USED AND  LIST THE EXTENTS
-------------------------------------------------------------------
SELECT T.DBNAME, T1.NAME, T.SPACE, T.PQTY*4 AS "PQTY",
       T.SQTY*4 AS "SQTY",
       (T.SPACE - (T.PQTY*4)) / (T.SQTY*4) AS "# EXT",
       (T.SPACE*100)/(T.PQTY*4) AS "% PQTY USED"
       FROM SYSIBM.SYSTABLEPART T, SYSIBM.SYSTABLESPACE T1
       WHERE T.DBNAME    =  'DSNDB06'
       AND   T1.DBNAME   = T.DBNAME
       AND   T1.CREATOR  = 'SYSIBM'
       AND   T.TSNAME    = T1.NAME
       AND   T.STORTYPE = 'E'
       AND   T.PQTY      > 0
       AND   T.SPACE     > 0
       ORDER BY T.DBNAME, T1.NAME;
-------------------------------------------------------------------
-- 2. FIND OUT THE LAST TIME RUNSTATS WAS RUN ON THE DATABASE
-------------------------------------------------------------------
     SELECT 'THE LAST RUNSTATS FOR THIS DB WAS: ', MAX(STATSTIME)
       FROM SYSIBM.SYSTABLESPACE
       WHERE DBNAME = 'XXXXXXXX';
-------------------------------------------------------------------

-- 3. FIND OUT THE NUMBER OF VARYING-LENGTH ROWS THAT HAVE BEEN
   -- REALLOCATED TO OTHER PAGES BECAUSE AN UPDATE.
   -- FARINDREF : IF THIS PARM IS LARGE, THIS INDICATES HIGH I/O ACTIVITY ON THE
   --             TABLESPACE. WE HAVE TO REORGANIZE THE TABLESPACE IF THIS PARM
   --             INCREASES OVER A PERIOD OF TIME.
   -- PERCACTIVE: IS THE MINIMUN % OF PAGES IN A TABLESPACE OR
   --             THE AMOUNT OF 'FREE-SPACE' CURRENTLY ON A PAGE.
   -- PCTPAGE   : IS THE CURRENT % OF UNUSED PAGES IN A TABLESPACE.
   -- NOTE      : THE 'PERCACTIVE %' SHOULD ALWAYS BE LESS THAN THE PCTPAGES %
   -------------------------------------------------------------------
SELECT A.DBNAME, B.TSNAME, B.CARD, B.PERCACTIVE, A.PCTPAGES,
       B.NEARINDREF, B.FARINDREF, B.PERCDROP
       FROM SYSIBM.SYSTABLES A, SYSIBM.SYSTABLEPART B
       WHERE A.DBNAME      = 'XXXXXXXX'
       AND   A.CREATOR     = 'AAAAAAAA'
       AND   A.TYPE        = 'T'
       AND   B.DBNAME      = A.DBNAME
       AND   B.TSNAME      = A.TSNAME
       ORDER BY A.DBNAME, B.TSNAME;


-------------------------------------------------------------------
-- 4. THIS QUERY MONITORS THE TABLESPACES CHARACTERISTICS AND THE CURRENT
--    STATUS, LOCK RULES, CLOSE RULES AND SEGMNENT SIZE.
-------------------------------------------------------------------
SELECT DBNAME, NAME, NTABLES, NACTIVE, SPACE, CLOSERULE, LOCKRULE,
       SEGSIZE, PGSIZE, STATUS
       FROM SYSIBM.SYSTABLESPACE
       WHERE DBNAME     =  'XXXXXXXX'
       AND   CREATOR    =  'AAAAAAAA'
       ORDER BY DBNAME, NAME;
-------------------------------------------------------------------
```

```
-- 5. THIS QUERY MONITORS THE SPACE ALLOCATED FOR EACH TABLESPACE (TS) AND
--     ALSO THE '%' OF SPACE SAVED IF THE TS IS USING THE 'COMPRESS' OPTION
--------------------------------------------------------------------------
SELECT T.DBNAME, T.NAME, T.BPOOL, T.LOCKRULE,
       T1.PQTY * 4 AS "PQTY", T1.SQTY * 4 AS "SQTY",
       T1.COMPRESS, T1.PAGESAVE, T1.FREEPAGE, T1.PCTFREE
       FROM SYSIBM.SYSTABLESPACE T, SYSIBM.SYSTABLEPART T1
       WHERE T.DBNAME    = 'XXXXXXXX'
       AND   T.CREATOR  = 'YYYYY'
       AND   T1.DBNAME   = T.DBNAME
       AND   T1.TSNAME   = T.NAME
       ORDER BY T.DBNAME, T.NAME;
--------------------------------------------------------------------------


-- 6. THIS QUERY MONITORS THE SPACE ALLOCATED AND THE SPACE USED FOR
--     FOR EACH TABLESPACE
--------------------------------------------------------------------------
SELECT T.DBNAME, T1.NAME, T.SPACE, T.PQTY*4 AS "PQTY",
       T.SQTY*4 AS "SQTY", (T.SPACE*100)/(T.PQTY*4) AS "% PQTY USED"
       FROM SYSIBM.SYSTABLEPART T, SYSIBM.SYSTABLESPACE T1
       WHERE T.DBNAME    =  'XXXXXXXX'
       AND   T1.DBNAME    = T.DBNAME
       AND   T1.CREATOR   =  'YYYYY'
       AND   T.TSNAME     = T1.NAME
       AND   T.STORTYPE  = 'I'
       AND   T.PQTY       > 0
       AND   T.SPACE      > 0
       ORDER BY T.DBNAME, T1.NAME;


-------------------------------------------------------------------------------------
--- 7. THIS QUERY MONITORS THE TOTAL SPACE ALLOCATED BY SECONDARY STORAGE FOR EACH TABLE
--- SPACE
-------------------------------------------------------------------------------------

SELECT T.DBNAME, T.NAME, T1.SPACE, T1.PQTY*4 AS "PQTY",
       T1.SQTY*4 AS "SQTY", (T1.SPACE - (T1.PQTY*4)) AS "% SQTY USED"
       FROM SYSIBM.SYSTABLESPACE T, SYSIBM.SYSTABLEPART T1
       WHERE T.DBNAME    =    'XXXXXXXX'
       AND    T.CREATOR  =    'YYYYY'
       AND   T1.DBNAME    = T.DBNAME
       AND   (T1.SPACE - (T1.PQTY*4)) > 0
       AND    T1.STORTYPE = 'I'
       AND   T1.PQTY       > 0
       AND   T1.SQTY       > 0
       AND   T1.TSNAME    = T.NAME
       ORDER BY T.DBNAME, T.NAME;
-----------------------------------------------------------------------
-- 8. THIS QUERY MONITORS THE TOTAL SPACE RESERVED THAT HAVE BEEN USED
--     FOR EACH TABLESPACE.
-----------------------------------------------------------------------
SELECT T.DBNAME, T.NAME, T1.CARD, T1.SPACE,
       (T.NACTIVE * T.PGSIZE)*100/T1.SPACE AS "% RESP USED",
        T.NACTIVE, T.PGSIZE
       FROM SYSIBM.SYSTABLESPACE T, SYSIBM.SYSTABLEPART T1
       WHERE T.DBNAME    =    'XXXXXXXX'
       AND    T.CREATOR  =    'YYYYY'
       AND   T1.DBNAME   = T.DBNAME
       AND   T1.TSNAME   = T.NAME
       AND   T1.SPACE     > 0
       ORDER BY T.DBNAME, T.NAME;
-----------------------------------------------------------------------
-- 9. THIS QUERY MONITORS THE TOTAL SPACE 'FREE' ON EACH TABLESPACE
-----------------------------------------------------------------------
SELECT T.DBNAME, T.NAME, T1.CARD, T1.SPACE, T1.PQTY*4 AS "PQTY",
       T1.SQTY*4 AS "SQTY", ((T1.SPACE*100)/(T1.PQTY*4)) AS "% FREE"
       FROM SYSIBM.SYSTABLESPACE T, SYSIBM.SYSTABLEPART T1
       WHERE T.DBNAME    =    'XXXXXXXX'
```

```
        AND    T1.DBNAME   = T.DBNAME
        AND    T1.TSNAME   = T.NAME
        AND    T1.PQTY      > 0
        AND (T1.SPACE/(T1.PQTY*4)) < 1
        AND    T1.SPACE     > 0
        ORDER BY T.DBNAME, T.NAME;
-----------------------------------------------------------------------
-- 10. THIS QUERY MONITORS ALL TABLESPACES THAT MUST BE REORGANIZED
-----------------------------------------------------------------------
SELECT DBNAME, TSNAME, CARD, PQTY*4 AS "PQTY", SQTY*4 AS "SQTY",
        NEARINDREF, FARINDREF, PERCDROP
        FROM SYSIBM.SYSTABLEPART
        WHERE DBNAME   =    'XXXXXXXX'
        AND    ((CARD > 0 AND (NEARINDREF + FARINDREF) * 100 /CARD > 0)
        OR     PERCDROP  > 0)
        ORDER BY DBNAME, TSNAME;


-----------------------------------------------------------------------
-- 11. THIS QUERY MONITORS THE TABLES AND THE CHARACTERISTICS REGARDING
--     CARDS, # OF PAGES USED, % OF PAGES USED, RECLENGTH.
-----------------------------------------------------------------------
SELECT DBNAME, TSNAME, NAME, CARD,
        RECLENGTH, KEYCOLUMNS, COLCOUNT, NPAGES, PCTPAGES
        FROM SYSIBM.SYSTABLES
        WHERE DBNAME    =    'XXXXXXXX'
        AND    CREATOR   =    'YYYYY'
        AND    TYPE       =   'T'
        ORDER BY DBNAME, TSNAME;
-----------------------------------------------------------------------
-- 12. THIS QUERY MONITORS THE TABLES THAT HAVE BEEN REALLOCATED
--     OUTSIDE OF THEIR ORIGINAL "PAGE".
-----------------------------------------------------------------------
SELECT DBNAME, TSNAME,
        ((NEARINDREF + FARINDREF) * 100) / CARD AS "PTTR",
        FARINDREF, NEARINDREF, PERCACTIVE
        FROM SYSIBM.SYSTABLEPART
        WHERE DBNAME    =    'XXXXXXXX'
        AND    CARD        > 0
        AND    (FARINDREF * 100) / CARD > 0
        ORDER BY DBNAME, TSNAME;
-----------------------------------------------------------------------
-- 13. THIS QUERY MONITORS THE INDEXES AND THE CHARACTERISTICS REGARDING
--     NLEVELS, CLUSTERRATIO, CLUSTERED =Y/N.
--     THE QUERY GETS ONLY THE CLUSTER INDEXES (CLUSTERING = YES).
-----------------------------------------------------------------------
SELECT DBNAME, TBNAME, INDEXSPACE,
        CLUSTERING||'-'||CLUSTERED AS "CLUSTER/SORTED",
        NLEVELS, FIRSTKEYCARD, FULLKEYCARD
        FROM SYSIBM.SYSINDEXES
        WHERE DBNAME   =    'XXXXXXXX'
        AND    CREATOR  =    'YYYYY'
        AND CLUSTERING  = 'Y'
        ORDER BY DBNAME, TBNAME, INDEXSPACE;
-----------------------------------------------------------------------
-- 14. THIS QUERY IDENTIFY INDEXES THAT SHOULD CONSIDER REORGANIZING
--                 USING THE *REORG* UTILITY.
-- LEAFDIST = AVERAGE DISTANCE BETWEEN SUCCESSIVE (LEAF PAGES) DURING
-- SEQUENTIAL ACCESS BY THE IXNAME.
-- IF THIS VALUE IS > 200 IS INDICATION OF A DISORGANIZED INDEX
-----------------------------------------------------------------------
SELECT DISTINCT B.DBNAME, B.TBNAME, A.IXNAME, A.LEAFDIST, A.FREEPAGE,
        A.NEAROFFPOS, A.FAROFFPOS
        FROM SYSIBM.SYSINDEXPART A, SYSIBM.SYSINDEXES B
        WHERE B.DBNAME   =    'XXXXXXXX'
        AND    B.CREATOR  =    'YYYYY'
        AND    A.IXCREATOR = B.CREATOR
        AND    A.IXNAME    = B.NAME
```

```
      AND LEAFDIST        > 0
      ORDER BY B.DBNAME, A.IXNAME;
----------------------------------------------------------------------
-- 15. THIS QUERY GETS INFORMATION ABOUT THE PGSIZE PARAMETER AND THE
--      SPACE ASSIGNED/ALLOCATED FOR EACH INDEX IN EACH TABLESPACE.
----------------------------------------------------------------------
 SELECT T.DBNAME, T.TSNAME, T.NAME, T.CARD, T1.INDEXSPACE,
        T2.SPACE, T2.PQTY*4 AS "PQTY", T2.SQTY * 4 AS "SQTY"
        FROM SYSIBM.SYSTABLES T, SYSIBM.SYSINDEXES T1,
           SYSIBM.SYSINDEXPART T2
        WHERE T.DBNAME   = 'XXXXXXXX'
        AND   T1.DBNAME  = T.DBNAME
        AND   T.CREATOR  = 'YYYYY'
        AND   T.TYPE     = 'T'
        AND   T1.CREATOR = T.CREATOR
        AND   T1.TBNAME  = T.NAME
        AND   T2.IXNAME  = T1.NAME
        AND   T2.IXCREATOR =  T1.CREATOR
        ORDER BY T.DBNAME, T.NAME, T1.INDEXSPACE;
----------------------------------------------------------------------
-- 16. THIS QUERY MONITORS THE TABLES & INDEXES AND IDENTIFIES CANDIDATES TO BE
--      ELIMINATED BECAUSE OF WASTED SPACE OR IDENTIFIES EMPTY TABLES.
----------------------------------------------------------------------
-- SELECT DISTINCT T.DBNAME, T.TSNAME,T.NAME,T1.INDEXSPACE,T1.NAME,
SELECT DISTINCT T.DBNAME, T.TSNAME,T.NAME,T1.NAME,
        T.CARD,T.NPAGES
        FROM SYSIBM.SYSTABLES T, SYSIBM.SYSINDEXES T1
        WHERE T.DBNAME    = 'XXXXXXXX'
        AND   T.CREATOR  = 'YYYYY'
        AND   T.TYPE     = 'T'
        AND   T1.DBNAME   = T.DBNAME
        AND   T1.CREATOR = T.CREATOR
        AND   T1.TBNAME  = T.NAME
        AND   T.NPAGES    < 6 ORDER BY T.DBNAME, T.TSNAME;
```

22.0   Grant and Revoke Privileges

```
SET CURRENT SQLID = 'HDBT99';

GRANT BIND,EXECUTE ON PACKAGE YYY.* TO XXXXXX;

GRANT BIND,EXECUTE ON PLAN YYYYYYY TO XXXXXX;

GRANT ALL ON YYY.* TO XXXXXX;

GRANT SELECT ON TABLE DB2T.ZZZZZZZZ TO XXXXXX;

GRANT SELECT ON TABLE DB2T.ZZZZZZZZ TO PUBLIC;

GRANT ALL ON TABLE DB2T.ZZZZZZZZ TO XXXXXX;

GRANT UPDATE ON TABLE DB2T.ZZZZZZZZ TO XXXXXX;

GRANT EXECUTE, BIND ON PACKAGE YYY.YYYYYYY TO PUBLIC;

GRANT SELECT
ON TABLE  DB2P.ACDVBASE,
          DB2P.ACREBASE,
          DB2P.ACSSBASE,
          DB2P.ACTCBASE,
          DB2P.BBDPBASE,
          DB2P.BLOMBASE,
TO XXXXXX;

GRANT SELECT ON TABLE DB2T.ZZZZZZZZ TO PUBLIC;

GRANT SELECT ON TABLE DB2T.ZZZZZZZZ TO PUBLIC AT ALL LOCATIONS;


REVOKE ALL PRIVILEGES
ON TABLE DB2T.SHDABASE
FROM XXXXXX;

REVOKE ALL ON TABLE DB2P.ZZZZZZZZ
FROM XXXXXX;

REVOKE ALL ON TABLE DB2P.ZZZZZZZZ
   FROM PUBLIC;
```

23.0 DB2 UDB AIX / UNIX

23.0.1 Clustering

The IBM SP Product Line represents Cluster Complexes of 1 to 1024 individual POWER, POWER2 or POWER PC nodes. An SP Cluster can also include 2 to 16 (32 under UDB ESE Version 8.1) way SMP Nodes (Symmetrical Multi Processor).

The Load Leveler queuing system and AIX Cluster support work in unison to make an SP Cluster function as a Unit. Networks (Front End = Ethernet, Backend = FDDI).

This type of configuration can manage concurrent logging of many users across many machines (up to a maximum of 45).

Single Clustering Criteria:-

1. One System image and one domain address per cluster.
2. Minimized Operation System and Command Mods.
3. The workload is within the limits of the HW/SW.
4. Open Systems and Standard Protocols
5. Heterogeneous architectures supported viable sensible.
6. System Network Boundaries masked from Base.
7. Addresses four Software Subsystems:-
   a. Domain Name.
   b. Password system.
   c. Configuration of Software.
   d. File System (Network only).

Cluster Members are made up of:- Applications Action Servers, File Servers, Password Servers and a Cross Reference System. Access to the Computer and Application Servers is provided via a front door Network (Ethernet). The Computer and Application Servers communicate with the file and password servers via a backdoor (FDDI Network) to isolate I/O Traffic from the Users.

The Cluster Consoles are connected to a set of terminal servers that reside on a secure subnet. A stand alone server on this subnet is used by operations and support staff to access the console of any machine using any Terminal. Locally developed software monitors and logs the console data from each Clustered Machine.

Only one System Image:- A single view of the Cluster File System hierarchy from any machine in the cluster can be achieved through the use of a remote file system (NFS or DCE). Each machine contains it's own copy of the Operating System and Application Set based on an automatic reference system. The reliability of servers can be improved through the use of HANF (High Availability Network File) on RS6000 systems and above.

Staff involvement would probably be less than supporting an equivalent mainframe system that carried a similar work load. But running costs would be less than a similar operation on the mainframe, not withstanding the up coming mainframe OS Z-Servers. The cyclic "leap froging" nature of competition, even from platforms developed by the same manufacturer, make dramatic gains of one platform versus another temporary at best and is usually limited to one or two years of advantage one way or the other.

24.0   SMIT

AIX's System Administration and Management Tool is called SMIT (**S**ystem **M**anagement **I**nterface **T**ool).

The "capabilities of SMIT are many and varied as follows:-

Software Installation, Physical and Logical Storage Devices, Users, Security, Workstation Installation, Management, Communications Applications, Services, Print Spooling (Printing), Problem Determination, Performance and Resource Scheduling, System Environments, Subsystems and Processes, Applications.

SMIT Dialog Symbols:- * = Entry Required, # = Numeric field, X = Hexadecimal Field, / = File Name Field, + = List Available, [] = Field Display (ASCII), Arrow Button = Fixed Set of Options (Motif display), List Button = List of Choices (Motif display).

SMIT Functions Keys:- F1 = Help, F2 = Refresh Screen, F3 = Cancel, F4 = Display Selection List, F5 = Undo entry - reset to default, F7 = Edit or Select, F8 = Display Fast Path Name, F9 = Shell escape, F10 = Exit SMIT, Enter = Execute, Tab/Shift Tab = Move between options.

 Fast Paths - SMIT allows you to bypass menu levels and enter a task directly through the use of a fast path name. Common Fast Path Names:- dev = Devices, diag = diagnostics, jfs = journal file system, nfs = network file system, sinstallp = software installs, spooler = Print Que. management, system = system management, tcpip = tcp/ip management, user = user administration.

To create a SMIT script file:

smit -s /tmp/smit.script - Creates a script file in /tmp

To create a SMIT log file:-

The location of the log file may be set using the smit -l <pathname> option.

smit -l /tmp/smit.log -vt - Use /tmp to hold log file.

SMIT verbose tracing:-

smit -vt

SMIT panels, command link descriptions, option defaults, and classes reside in the :-

/usr/lib/objrepos directory.

SMIT is started by:-

smitty   - Start SMIT at the top-level-menu

smit user - Start SMIT at the "user-admin" level submenu.

SMIT allows you to take a "test drive" utilizing all of it's features.

Invoke SMIT with # smit -x to get a feel for how it operates (the equivalent to having a test drive in a fast car!).

You may use PF6 within SMIT to display the AIX command and arguments it will execute before doing (committing) the update.

SMIT can be displayed in Motif or ASCII

SMIT creates an Audit Log for each SMIT session for each User.

Together with the Log File, SMIT appends the AIX commands invoked during the session to a local $HOME/smit.script file. The script file can be duplicated for example changing the device name for all the devices you want to add (Yank and Put (YYP - this copies a line to the space below it in vi)) then alter as necessary.

SMIT lets you bypass menu levels and enter a task directly using SMIT Fast Path.

Fast path command algorithms:- mk, ls, ch, rm

Objects:- dev, user, vg, tty, diskette, tape etc.

There are three type of SMIT panels:- Menu - List Task Options,
Selector - Request additional input, Dialog - request values
for command arguments and options.

DSMIT - Distributed System Manager Interface tool. Installed
separately. Machines managed by DSMIT are grouped into domains.
Subsets of machines within a domain are
called the 'working collective' (for want of a better phrase!).
DSMIT uses MIT Kerberos Security Services for V5 and upwards.

Invoke DSMIT by entering SMIT=d

24.0.1 VSM - Visual System Management -

For those who like glitz, VSM is the tool to use. VSM is an
object-oriented GUI (Graphical User Interface) that allows you
to perform system administration and Management tasks by a
mouse driven, point and click, drag and drop X-based Desktop
environment. To perform any option simply drag
the desired object icon onto the task icon.

```
VSM Commands:- $ installm - Installation Manager
               $ xmaintm - Software maintenance
               $ devicem  - device manager
               $ xprintm  - print manger
```

```
Backups:-
               # mkszfile
               # chdev -1 rmt0 -a block_size=512
               # tct1 -f /dev/rmt0 rewind
               # mksysb /dev/rmt0
               # chdev -1 rmt0 -a block_size=<blocksize>
```

NIM - Network Installation Manager - replaces Diskless
Workstation Manager (DWM)for administering diskless/dataless
clients and stand alone work stations.

If you want to see all the environment variables:-

```
               MANPATH=/afs/Austin/common/usr/man

               Lang=En_US

               LOGIN=xxxxxx
```

To repeat the previous command type !!

Other Examples:-

```
rm * .tmp
remove temporary files in directory

cc * .c
compile all C programs in directory

Cat prog(123)..c
Catalog certain programs

ls (adt)*
Look at all files or
directories starting with
a, d or  t
```

25.0   ROOT

What is a "root"??

The word "root" in AIX means two things: 1. The root user is the "super user" (su) the root directory is the top of the directory tree.

The system comes with one user, the root. The root user also has a password. The root user can read / write into any directory. Only the root user can change operating system characteristics.

To change your user Id., to the root, use the su (superuser) command. To go back to your Id., from the root press <crtl-d>. If you can't remember which ID you are currently using type *whoami*.

The top of the file system at "/" is known as the root directory. Picture an inverted B Tree (a tree upside down).

Sticky Bit (seldom used now)

"Sticky bit" attribute changes the way the link permission acts. Sticky bit was used to request that a program stick around in memory after being accessed, thus perhaps saving unnecessary I/O if invoked repeatedly.

To view permissions on a file, on the command line type **ls -l** and you will see something like this:- -rw-r—r-- 1 mark system 41212 July 8 20:15 -rw-r—r-- 1 mark system …..

rw files have read write permission for owner Mark.

Group = system

r-- = read permission for others.

To change permissions use the **chmod** command (eg:- chmod 755 file-name).

The **unmask** command sets the default for newly created files.

The **chown** command changes ownership of a file (only the root can do this).

<u>25.0.1 Various Processes</u>

ps = Process Status, kill = terminate, & = execute in the
    background, <ctrl z> = Pause

fg = put command in foreground.

To get a list of processes running = > ps, > ps -1 = long status

Note:- Do not use KILL -9 when wanting to STOP a DB2 Process.
Instead use **db2stop force**. Using Kill -9 May bring the whole
instance down!!!

26.0 DB2 UDB EEE (a.k.a. as ESE from Version 8.1 Up)

Note:- EEE = Enterprise Edition Extended.
        ESE = Enterprise Server Edition
        In Version 8.1, EE and EEE have been merged to
        form ESE.

26.0.1 Unix ESE Installation

With Unix you can choose the install method you want to use. The available methods are:-

1. The DB2 Installer or DB2Setup program. This program is supplied with the DB2 UDB product and DB2Setup is the recommended choice. It is most like the set up program for Windows NT or OS/2 in that it detects and configures communication protocols to use with the DAS and DB2 instances. The other methods require you to manually configure your system.

2. The install command found in your UNIX environment. For example in AIX the install command will install the DB2 UDB products. But using the UNIX command requires you to know the complete syntax.

3. There may be an installation utility available in your UNIX environment. For Example in AIX the System Management Interface Tool (SMIT) is available to install software. This is a better choice than using the install command. However there is no communication detection and configuration performed.
   Also the DAS and DB2 instance are not created.

   The DB2 installer program is the recommended choice. It will also be consistent across all UNIX environments. To run the DB2 Installer program, perform the following steps:-

        a. Log on to the system as the root (Unix System Administrator).

        b. Mount the CD-ROM file system.

       c. From the directory where the CD-ROM is
         mounted, type db2setup. The following screen
         appears:-

       Customize selection:-

       DB2 Administration Client
       DB2 UBD Enterprise Edition
       DB2 Connect Enterprise Edition
       DB2 Software Developer's Kit

The DB2 installer first scans your system to check for required
or prerequisite software products. Notice that you can select
products and their components. You must select the products you
want to install. You may select CUSTOMIZE for each product you
want to install. The (*) denotes your choices.

Some components are required while others are optional. An *
next to an optional component means that the component will be
installed. By default the following options will be installed
unless deselected:-

1. Java Support
2. Replication
3. Code page conversion support.

After selecting the components to install you will have the
option of creating the default environment. You can create a
DB2 instance and a DB2 Administration Server (DAS)
Instance. They may also be created after installation. However,
creating the instances now will automatically configure them
for communication with DB2 UDB.

To create a DB2 or DAS instance in UNIX you need a user name
and a group name. You may change the defaults of db2instl for
user name and db2instl1 for group name.

Select Properties to view or change more options. Now you can
specify the authentication type for the DB2 Instance (e.g.
Server, Client, DCS, DCE, etc) put an * in
the right slot. Notice, you have the ability to create the
SAMPLE database and automatically start DB2 when DB2 is booted.
From here you can customize protocols that will be used with
DB2 UDB. TCP/IP is usually selected with a *.

Select Properties to view or change the default configuration.

You can also create a user that can run user-defined functions.

For security purposes it is recommended that you use a User ID in which <u>fenced</u> UDFs will execute different than those of the DB2 instance and DAS instance. Running UDFs under this ID will control the scope and authority of the actions that can be performed by the function.

The DB2 Administration Server instance is a special DB2 instance that enables remote administration of a DB2 UDB server. You may set up a User ID, Group ID and password for the DAS Instance or you may use the defaults.

The Summary Report screen is displayed when you have completed selecting the DB2 UDB components to install.

After the messages have been displayed, exiting out of the installation screen will complete the UNIX installation process.

In some UNIX operating systems you may have to modify the Kernel configuration parameters to install and run DB2 UDB. Check "The Quick Beginnings" User Guide for UNIX.

The directory for installation of the DB2 UDB products will vary depending on the UNIX environment. For example in AIX, DB2 UDB products are installed in the /usr/lpp/db2_vv_rr directory where vv = the version and rr = the release.

The directory structure created under /usr/lpp/db2_06_01 will be similar to the following:-

```
Readme        -   Read me file
Adm           -   System administrator and executable files
Adsm          -   ADSTAR distributed storage manager files
Bin           -   Binary executable files
Bnd           -   Bind files for utilities
Cc            -   Control Center files
Cfg           -   Default system configuration files
Doc           -   On-line books
Function      -   Default location for UDFs
```

```
Function/        -   Default location for unfenced UDFs
unfenced
Install          -   Installation program
Instance         -   Instance scripts
Java             -   Java programs
Lib              -   Libraries
Map              -   Map files for DB2 connect
Misc             -   DB2 tools and utilities
Msg/$L           -   Message Catalogs
Samples          -   Sample programs and sample scripts
```

To register DB2 using UNIX operating native installation tools you must use the db2licm command to register DB2 UDB:-

db2licm -a /cdrom/db2/license/filename.lic

filename.lic represents the file containing valid license information.

It is possible to install different versions of DB2 UDB products on the same UNIX server because the product files will be installed in a different path.

TO ACCESS AIX:-

```
    1. OPEN A TELNET SESSION
    2. CONNECT TO THE INSTANCE
    3. LOGIN: JOHNDOE
    4. P/W:- XXXXXXXX
    5. CONNECT TO THE DATABASE:-
```

DB2 CONNECT TO *database* USER johndoe USING xxxxxxxx

27.0   UNIX Scripts

1. Calculate sizes of UDB Tablespace and Indexspace. This script will calculate the sizes of UDB Tablespace and Indexspace files for filesystem tablespaces dms and sms.

2. If first parm = "?", then display usage syntax and examples.

3. This script is used with list alias to pipe ls output to '*more*'.

4. Reorg Script. Reorg.ksh <instance> < dbname> <reorg sql - optional> <output file optional>

5. UDB Offline Backup Utility. This script is called by the backup_db (the controlling script). This script runs the Offline database backup utility. It first looks to see if there are any connections to the database. If there are, it forces them off. The database is deactivated, on Offline backup is taken, then the Database is activated.

27.0.1 Script 1 - Calculation

```
#! /bin/ksh
# This script will calculate the sizes of udb tablespace
# and indexspace
# files for filesystem tablespaces (dms or sms)
#
# check for 2 input variables. If not, exit with a help
# message
#
if  [ $ {#} -lt 2 ]
then echo  "calc.dbspace.ksh <instance name> <database name>"
     echo
     echo  "output file is dbspace_<instance name>_<database name>.
txt
     echo
     exit
fi


DB2INSTANCE='echo $1 |  tr " [ :upper: ] "  " [ : lower : ] "
\

UDBNAME= 'echo $2  |  tr " [ :upper: ] "  " [ :lower : ] "'

UTLSCRDIR="/DBA/Utility"
Export UTLSCRDIR
#echo $UTLSCRDIR

if [  !  -f $UTLSCRDIR/${DB2INSTANCE}.txt  ]
then
     echo $ {DB2INSTANCE} > $UTLSCRDIR/$ {DB2INSTANCE}.txt
fi

DFILSYS="db2db"
LFILSYS="db2log"
find / $ { DFILSYS / ${ DB2INSTANCE } / $ ( UDBNAME } /* -size +1c –ls | awk '$3
= = "-rw-------" {print $11, $7; sum += $7 } END  {print "\nTotal for tablespaces and
Indexes:", sum, "(bytes), ",int(sm / 1024), "(kb), " , sum /1024/1024, " (mb), " sum /1
024/1024/1024, " (Gb) "  } ' > dbspace_$ {DB2INSTANCE }_$ {UDBNAME }.txt

echo ' ' >> dbspace_$ { DB2INSTANCE }_${UDBNAME}.txt

find /$ { LFILSYS } / $ { DB2INSTANCE } / $ { UDBNAME } / * -size +lc –ls –print |
awk '$3
```

```
= = "-rw-------" {print $11, $7; sum += $7 END   { print "\nTotal for Logs: ",sum. "
(bytes),",
int (sum /1024), " (kb), ", sum /1024/1024, " (mb) , ", sum /1024/1024/1024, " (Gb)
' } ' >> dbspace_$ {DB2INSTANCE}_${UDBNAME.txt

echo ' ' >> dbspace_$ {DB2INSTANCE }_$ { UDBNAME } .txt

find / $ { LFILSYS } / $ { DB2INSTANCE } / $ {UDBNAME } / * -size +lc -ls -print |
awk '$3
= = "-rw-------" {print $11, $7; sum += $7 } END { print "\nTotal for Logs: ", sum, " (bytes)
,
" ,int (sum / 1024), " (kb), " , sum /1024/1024, " (mb), ", sum /1024/1024/1024, "(Gb) " }
' >> dbspace_$ {DB2INSTANCE}_$ { UDBNAME }.txt
echo
echo "***********************************************************************
**"
echo "Output is in: " $ { PWD } " /dbspace_" $ {DB2INSTANCE } "_" $ {UDBNAME } ".txt"
echo
echo "Total space used is at the bottom of this file"
echo 'Total ' $ {PWD } / dbspace_$ {DB2INSTANCE }_$ {UDBNAME ) .txt
exit
```

27.0.2 Script 2 - Unix Usage Display

```
#!/bin/ksh
#
#!/bin/ksh –vx
#-------------------------------------------------------------------------------------------------
# if 1st parm = "?" then display usage syntax and examples
#-------------------------------------------------------------------------------------------------
function helpmsg
{
echo
echo   "Usage:      findid <id/string> "
echo   "Examples: findid a123456
echo   "              findid lastname (could be useful to find someone's id) "
echo
}   # # end of function helpmsg

userid = $1
if [  "$1  ]
then
    if [ "$1  = " ? "
    then
        helpsmg  # # see above functon
```

```
      exit
   fi ## [ $1 = " ? " ]
#
# Prompt for id/string if not provided
#-------------------------------------------------------------------------------------
---
if [ ${#} -lt 1 ]
then
    echo
    echo " Correct syntax: findid [ id/string ] "
    echo "Use ' findid  ? ' for help "
    echo " or you can follow the prompts below: "
    echo
    echo " Enter id/string to search /etc/passwd, /etc/group & ypcat passwd "
    read INPUTID
    userid= $INPUTID
fi ## [ ${#} –lt 1 ]

echo " Searching for id/string ($userid) via 'grep –i $userid /etc/passwd' "
more /etc/passwd | grep -i $userid
echo
echo " Searching for id/string ($userid) via 'ypcat passwd | grep -i $userid' "
ypcat passwd | grep -I $userid
echo
echo "Searching for groups containing ($userid) via 'grep -i $userid /etc/group'
"

grep –i $userid /etc/group | awk ' BEGIN { FS=":" } { print $1 } '

exit
```

27.0.3 Script 3 - Pipe to output

```
#! /bin/ksh
#
# used with list alias to pipe ls output to more
# defaults to "list * " -→ "ls –lApd * | more "
#
if [ ${#} –gt 0 ]
then
    if [ $1 == " ? "
      then
      echo
        echo " used with list alias to pipe ls output to more "
        echo 'list = ls –lApd $* | more '
```

```
        echo  "USAGE:     list  <file  name/pattern>"
        echo  "Example: list  *.java "
        echo  "Default: list * "
        echo
        echo  "Note: assumes alias list='/dba/Utility/list.ksh' "
        echo
        exit 1
fi
fi
if [  $ { # } –lt  1  ]
then
    ls  =  -lApd *  | more
if [  $ { # }  -gt  0  ]
then
    if [  $1 = =  " ? " ]
    then
        echo
        echo " used with list alias to pipe ls output to more "
        echo 'list   = ls  -lApd  $ *  |  more '
        echo "USAGE:  list  <file name/pattern> "
        echo "Example: list *.java"
        echo "Default:  list  * "
        echo
        echo " Note: assumes alias list = ' /dba/Utility/list.ksh' "
        exit 1
    fi
fi
if [  $  { # }  -lt  1  ]
then
    ls -lApd  *  | more
else
    ls -lApd  $ *  | more
fi
exit 0
```

27.0.4 Script 4 - UNIX Reorg

```
#!/bin/ksh
#
# reorg script
#
# USAGE:  reorg.ksh <instance> <dbname> <reorg sql – optional> <output file – optional>
#
#              reorg.ksh db2dw01 infatest
```

```
#
#
typeset  DB2INSTANCE=${1-DB2INST}
typeset  dbname=${2-sample}
typeset  reorgsql = ${3 -$PWD/reorg.sql}
typeset  outfs = ${4-/dba/${DB2INSTANCE}/${dbname}/Reorg_Logs}

. /export/home/$DB2INSTANCE} / sqllib/db2profile

if [ ! -d $outfs ]
then
    mkdir  $outfs
fi


UTLSCRDIR = "/DBA/Utility"
Export UTLSCRDIR
#echo  $UTLSCRDIR

if [ ! -f $UTLSCRDIR/$ {DB2INSTANCE}.txt ]
then
    echo  $ {DB2INSTANCE} > $UTLSCRDIR / $ {DB2INSTANCE}.txt
fi

MAILDIR="/dba/Mail"
export MALDIR
#echo $MAILDIR

MAILIDS='paste –s ${MAILDIR}/mailto_DBA.dat'
export MAILIDS
#echo  $MAILIDS

YMD='date  +%Y%m%d_%T
i = 1

db2  -tvf $PWD/reorg.sql > $outfs / $dbname.reorg.$YMD.txt

export MAILDIR
#echo $MAILDIR

MAILIDS='paste –s ${MAILDIR} / mailto_DBA.dat'
export MAILIDS
#echo  $MAILIDS

YMD= 'date +%Y%m%d_%T
i = 1
```

```ksh
db2 –tvf $PWD/reorg.sql  >  $outfs/$dbname.reorg.$YMD.txt

if [  $?  !  = 0  ]
then
    cat $outfs / $dname.reorg.$YMD.txt |  mailx –s "ERROR IN REORG $ {dbname }"
$MAILIDS

    /dba/ONCALL / email_oncall.csh  $outfs / $dbname.reorg.$YMD.txt $UTLSCRDIR/
${DB2INSTANCE}.txt
fi

find ${outfs} -name "$dbname.reorg*" -mtime +60 –exec rm –f {} \;

exit
```

27.0.5   Script 5 – Save a backup

```ksh
#!/bin/ksh
#
# Save a backup of the file prior to editing, into the same directory
# where the original file exists and add a suffix of .vibackup and datetime
#
for file in $*  ; do
   if [ [  -a $file  ] ]  ; then
     cp $file ${file}.vibackup.'date +%Y%m%d+_T '
   fi
done
/usr/bin/vi  $ *
```

28.0   Sample Java Script

```java
import   java.io..*;
import java.lang.*;
import java.sql.*;
import java.util.*;
import java.util.StringTokenizer;

//  CCXAnnPost ST_HIST table delete, using result set

public class CCXAnnStHist {

//      private String iCntrc_id;
//      private String iCntrc_sfx_cd;
//      private String iSrc_sys_cd;
//      private int iSeq_num_int;
//      private int iMetcnct_cust_id_int;
        private Connection con;


        public static void main(String[] argv) throws IOException{
            String db                   = " ";
            int recTotal                = 0;
            int insCount                = 0;
            int delCount                = 0;
            int insPol                  = 0;
            int delPol                  = 0;
            String stmtNM               = "";


            String url                  = "jdbc:db2:E8412IO4";
            String userName             = "e8412dba";
            String password             = "e8412dba";
            db                          = "IO4";
/*
            String url                  = "jdbc:db2:E8412IOL";
            String userName             = "E8412DBA";
            String password             = "bryan";
            db                          = "IOL";
*/
/*
            String url                  = "jdbc:db2:E8412QOL";
            String userName             = "E8412DBA";
            String password             = "OLCIUDB";
            db                          = "QOL";
*/

/*          String url                  = "jdbc:db2:E8412POL";
            String userName             = "E8412DBA";
            String password             = "OLCIUDB";
            db                          = "POL";
*/
```

```java
                try{
//                      establish connection
                        Class.forName("COM.ibm.db2.jdbc.app.DB2Driver");
                        Connection con =
                        DB2DRIVER.getConnection(url,userName,password);
                        con.setAutoCommit(false);
                        System.out.println("Connection to "+db+" made");


                        PreparedStatement psStHistSEL = con.prepareStatement(
                        "SELECT METCNCT_CUST_ID, CNTRC_ID, CNTRC_SFX_CD,
                        SRC_SYS_CD, SEQ_NUM FROM E8412DBA.TOLST_HIST T0 "+
                        "WHERE   NOT EXISTS (SELECT 1 FROM
                        E8412DBA.TOLCCX_OWN T1 WHERE T0.METCNCT_CUST_ID =
                        T1.METCNCT_CUST_ID AND T0.CNTRC_ID = T1.CNTRC_ID AND
                        T0.CNTRC_SFX_CD = T1.CNTRC_SFX_CD  AND T0.SRC_SYS_CD =
                        T1.SRC_SYS_CD AND T0.SEQ_NUM = T1.SEQ_NUM ) "+"
                        AND EXISTS (SELECT 1 FROM  E8412DBA.TOLCCX_OWN
                        T1 WHERE T0.CNTRC_ID = T1.CNTRC_ID AND T0.CNTRC_SFX_CD
                        = T1.CNTRC_SFX_CD  AND T0.SRC_SYS_CD = T1.SRC_SYS_CD
                        AND T0.SEQ_NUM = T1.SEQ_NUM )" );

                        PreparedStatement psStHist    = con.prepareStatement(
                        "DELETE FROM E8412DBA.TOLST_HIST AS T0 "+"
                        WHERE   T0.CNTRC_ID = ?
                        AND T0.CNTRC_SFX_CD = ?
                        AND T0.SRC_SYS_CD = ? AND      T0.SEQ_NUM = ? AND
                        T0.METCNCT_CUST_ID = ? ");


/*                      String icntrc_id        = " ";
                        String icntrc_sfx_cd    = " ";
                        String iseq_num         = " ";
                        String isrc_sys_cd      = " ";
                        String imetcnct_cust_id = " ";
*/
                        int iseq_num_int   = 0;
                        int imetcnct_cust_id_int = 0;

//              while ((recin  = rin.readLine())!= null){
//                      recTotal += 1;
//                              System.out.println("recin  = "+recin);


                        try{

                        stmtNM  = " selCustCntrcNM ";
                        ResultSet rsCount  = psStHistSEL.executeQuery();
                        while (rsCount.next()) {
                        String imetcnct_cust_id = rsCount.getString(1);
                        String icntrc_id  = rsCount.getString(2);
                        String icntrc_sfx_cd    = rsCount.getString(3);
                        String isrc_sys_cd      = rsCount.getString(4);
                        String iseq_num         = rsCount.getString(5);
```

```java
                iseq_num_int            = Integer.parseInt(iseq_num);
                imetcnct_cust_id_int    = Integer.parseInt(imetcnct_cust_id);


//      System.out.println("cntrc_id="+icntrc_id+" sfx="+icntrc_sfx_cd+"
src="+isrc_sys_cd+" seq="+iseq_num+" MC="+imetcnct_cust_id);

                    psStHist.setString (1, icntrc_id);
                    psStHist.setString (2, icntrc_sfx_cd);
                    psStHist.setString (3, isrc_sys_cd);
                    psStHist.setInt    (4, iseq_num_int);
                    psStHist.setInt    (5, imetcnct_cust_id_int);

                    stmtNM  = " insCustCntrcNM ";
                    delCount  = psStHist.executeUpdate();
                    delPol   += delCount;
                    long ts25 = System.currentTimeMillis();
                    System.out.println("finished: "+stmtNM+icntrc_id+ts25);

                            } //end-while


//                  long ts25 = System.currentTimeMillis();
//                  System.out.println("finished: "+stmtNM+icntrc_id+ts25);



                    } //end-try
                    catch (SQLException sqle){
                    while (sqle != null){
//                  System.out.println("SQL exception on contract:
                    "+icntrc_id+stmtNM+sqle.getMessage());
                    System.out.println("SQL exception on contract:
                    "+stmtNM+sqle.getMessage());

                    sqle=sqle.getNextException();
                    }   //end-while
                    }   //end-catch
                    con.commit();
//          }   //end-while
//              rin.close();
                con.commit();
            System.out.println("Total Records Processed: "+recTotal);
            }   //end-try
            catch (SQLException sqle){
                    while (sqle != null){
            System.out.println("SQL exception: "+stmtNM+sqle.getMessage());
                                sqle=sqle.getNextException();
                    }
            }
//          final
//          catch (EOFException e){
//          System.err.println("End of file");
//              }
```

```
            catch (ClassNotFoundException cnfe){
            System.err.println("DB2 driver class not found");
            }
        }


    }
```

29.0   EEE Selected Functions Overview

Function	Description
BACKUP	Make a copy of the database in whole or in part. 1. The name of the database to be backed up. 2. The names of the tablespaces to be backed up. 3. Is backup offline or online – default is offline. 4. The name of the devices or directories in which the backup files are to be created. Example:- BACKUP DATABASE accounts 　　　　　TABLESPACE Userspace1 ONLINE 　　　　　To d:\backups If you add containers, do a backup before adding containers.
db2admin	The administration server is created automatically when you install UDB, and it is started automatically whenever you reboot your system. db2admin start – manually starts the administration server. db2admin stop – manually stops the administration server. db2admin – displays the instance name of the Administration Server.
db2batch	This command invokes a tool that is useful for Performance Measurement. Example:- db2batch –d testdb –f infile.sql –r outfile.txt the above command executes statements in the input file 'infile.sql' and captures the results in the output text file 'outfile.txt'. The database referenced is 'testdb'.
db2bfd	Examines the content of a bind file, produced by a PREP command with the bind option:- db2bfd –b –s –v accts.bnd The above command displays the contents of the bind file named accts. bnd including it's bind options, SQL Statements and host variable declarations.
db2gov	This command invokes a utility called the DB2 Governor, which enables you to place limits on resources consumed by certain users or applications. db2gov start testdb config.txt govlog.txt The above command starts the Governor in database testdb in configuration file config.txt and logs the actions in file govlog.txt

Db2rbind	Rebinds all packages in a database. Use after reorg. db2rbind company /1 rebind.log The above command rebinds all the packages in the company database And directs any error messages to a file named rebind.log

30.0　How to Create an Event Monitor

1.　db2 "create event monitor CHKCOST for statements write to file '/tmp/EVMON'"
2.　mkdir /tmp/EVMOUT
3.　chmod 775 /tmp/EVMOUT
4.　db2 "set event monitor CHKCOST state 1" (to activate the event monitor)

To read the collected information from the event monitor output file under the directory /tmp/EVMON type:-
　　　　Db2evmon -path /tmp/EVMOUT

Pay particular attention to "sort overflows" and total sorts (to obtain % sort overflow), total sort time as well as user and system CPU times.

Also, analyze bufferpools by obtaining snapshot for the bufferpool(s). From this calculate the overall bufferpool hit ration using the following formula:-

1 -((bufferpool data physical reads + bufferpool index physical reads) /(bufferpool data logical reads + bufferpool index logical reads).

A low bufferpool hit ratio means that the bufferpool should be increased or better yet assigning types of table spaces to their own bufferpools (i.e. tempspace, randomly accessed tables, etc.).

Note:- For a particular database, try altereing the IBMDEFAULTBP to NPAGES -1 (so BUFFPAGE can be utilized) and increase the database configuration parameter BUFFPAGE to 30,000 (4K) pages. After doing this, the bufferpool hit ratio should increase by 30% or more.

31.0   Example of a User Defined Function

Introduction

In IBM's DB2 UDB there are four types of functions that are available for use in SQL statements as follows:-

1. Built in functions – These functions are built into the code of the UDB System and are found in the SYSIBM SCHEMA, and include:

   Arithmetic and string operators: +, -, *, /, ||

   Scalar functions: substr, concat, length, days

   Column functions: avg, count, min, max, stdev, sum, variance

   In addition to the built-in functions in the SYSIBM schema, many other functions are shipped with UDB in the SYSFUN schema.

2. System generated functions – These functions are automatically generated when a distinct type is created. System generated functions include casting functions and comparison operators for the distinct type.

3. User Defined Functions - Sourced Functions – A sourced function duplicates the semantics of another function, called it's *source function.* Sourced functions can be very useful in allowing a distinct type to selectively inherit the semantics of it's source type. A sourced function can be an operator, scalar function or column function.

4. User Defined Functions – External Scalar Functions – An external scalar function is a function that is written by a user in a host programming language and returns a scalar value. External scalar functions are mostly written in Java or C. The CREATE FUNCTION statement for an external scalar function tells the system where to find the code that implements the function. An external scalar function may not contain any SQL statements. In other words, an

external scalar function may perform any computation you like on the parameters that are passed to it but it may not access or modify the database. The Hypergeometric Distribution for Estimating CPR Envelope Function (a.k.a. Hypertherotical or Htheor) is an example of a User Defined External Scalar Function described here in detail.

5. User Defined Functions – External Table Functions – A User Defined Function can return a table rather than a scalar value. Like an external scalar function, an external table function is written by a user in C or Java, and may not contain any embedded SQL statements. The program that implements an external table function must return one row of the result table each time it is invoked, and must indicate the end of the result table by a special return code.

Authorization

The creation of a user defined function requires SYSADM, DBADM, CREATEIN or IMPLICIT_SCHEMA authority on the database if the schema does not yet exist. In addition, if the function has the NOT FENCED
property a database-level authority called CREATE_NOT_ FENCED is required.

Register

The user must register the external function by executing a CREATE FUNCTION statement in each database where the function will be used.

Note:- Regardless of your operating platform, you should be very careful to protect the executable file that implements your external function against
accidental or unauthorized tampering. The database does not protect this file and will execute it whenever your function is invoked. If your function
implementation is replaced by another executable file, it could potentially do something harmful.

Function Parameters

This section will cover only the parameters used by the Hypertheoretical User Defined External Scalar Function. The

format of the create statement will be shown first, followed
by an explanation of each parameter used.

The Create Function statement for the Hypertheoretical
function follows:-

```
CREATE FUNCTION schema name.Htheor(integer, float, integer, float)
          RETURNS float
          NOT VARIANT
          EXTERNAL NAME 'hgudf!Htheor'
          NO EXTERNAL ACTION
          LANGUAGE C
          PARAMETER STYLE DB2SQL
          NO SQL;
```

The parameters of the CREATE FUNCTION statement above
are for an external scalar function as pertaining to the
Hypertheoretical (a.k.a. Htheor) and are each described
separately below:-

1. <u>CREATE FUNCTION schema name.Htheor(integer, float,
   Integer, float)</u>

   The function is created in the SYSIBM.SYSFUNCTIONS
   Table under 'schema name.Htheor'. This is the name
   under which it is registered on the database. If for
   some reason the name needs to be changed at a later
   date, then the function will have to be dropped, the
   name changed and then the function re-created.

   The (integer, float, integer, float) are the
   parameters representing the data types of the columns
   identified when the function is invoked. Up to ninety
   data types may be used as parameters, dependent on
   the type of function invoked. In the Htheor function
   there are four input parameters.

   Note:- There is <u>not</u> a one for one relationship
   between the SQL and C data types. Thus in the case of
   function Htheor the (integer, float, integer, float)
   SQL datatypes will be translated by C and input to
   the main C program hgtuf as (long, double, long,
   double).

2. RETURNS FLOAT

This parameter specifies the data type that is the result of
your function's execution. Of course, since the external
function Htheor is implemented by a C program the actual
value returned by the C program is a two step process,
however in the case of our function, the C program hgudf
returns a data type which is float and the SQL result is also
a float data type.

3. NOT VARIANT

You must declare your function VARIANT or NOT VARIANT. VARIANT
means that your function might return different results from
two successive calls using the same parameters. An example
of a variant function would be a random number generator such
as the system provided function *rand,* which returns a random
floating point number between 0 and 1.

4. EXTERNAL NAME

This parameter identifies the function as an external function
and tells the system how to find the C function that serves it's
implementation. This C function must be compiled, linked and
placed in a directory on the server machine from where it can
be dynamically loaded by the database system when needed.

The most complete form of an EXTERNAL parameter gives the full
path name of the binary file that implements the function,
followed by a "!", followed by the name of the proper entry point
in that file. For example:- 'xya/dev/fin/ppa/bin/hgudf!Htheor'

The clause above tells the system to look in the file:- xya/dev/
fin/ppa/bin where the entry point is hgudf, which of course is
the highest level C program we want entered. If no path name
is specified the system looks for the function in the sqllib/
function directory associated with the database. EXTERNAL NAME
hgudf!Htheor assumes that the executable program hgudf has
been moved to the sqllib/function library, otherwise the full
path name is necessary.

5.   NO EXTERNAL ACTION

This parameter is mandatory, and it specifies whether the function performs some action that that affects the 'world' outside the database. For example, you might want to write a function that sends mail to someone, writes to a file or sets off an alarm. Of course you want the number of invocations of such a function to be predictable for example, if the function is used once in a SELECT list it should be invoked exactly once for each row returned by the query. The EXTERNAL ACTION parameter alerts the database optimizer to the existence of these functions, so that it will not modify the query in any way that changes the number of calls to the function.

6.   LANGUAGE C

The LANGUAGE parameter is mandatory, and specifies the programming language in which the function is implemented. This determines the linkage convention used by UDB for invoking the function. The options are C, JAVA and OLE, in our case it is C.

7.   PARAMETER STYLE DB2SQL

This parameter is also mandatory and identifies the conventions that are used for passing parameters to the external function. These conventions deal with issues such as how NULL values are represented, and how error conditions are reported. With PARAMETER STYLE DB2SQL the language must be C or OLE, and with JAVA you must specify the PARAMETER STYLE as DB2GENERAL.

8.   NO SQL

This mandatory parameter specifies that the external function Htheor contains no SQL statements.

Function Installation

The file containing the C program that implements your function may have any name you like and may contain more than one function body. The C program can include header files such as <stdio.h> and <math.h> and call standard C functions. Before UDB can use your external function, you must compile and link it and put the executable file in the appropriate directory on the database server.

The first step in installing your function is to create a *module definition file* that lists the function and all the entry points and modules. This file serves as input to the linker. Some sample 'module definition files' can be found in directory:- sqllib/samples/c Example:- On AIX / UNIX the module definition file is called an export file. If you link the hgudf.c file using the IBM XLC compiler you will need an export file named hgudf. exp with the following content:-

    #! export file for Htheor

Compile and link the file containing the program that implements your function. In the directory sqllib/samples/c you will find a script that you can use for this purpose. In AIX you can use the IBM file bldvaudf.bat. Before using this file (bldvaudf.bat) you should read it and modify it if necessary.

Place the executable file into the appropriate directory on the server machine. After linking program hgudf.c, the executable file is simply named hgudf. The following modules must be included in the link edit: gammln.c, factln.c, hypergeo.c and Htheor.c.   By default this executable should be placed in the directory sqllib/function on the server machine. If this is done the CREATE FUNCTION statement need not include a path name. Alternatively, your CREATE FUNCTION statement can specify a full path name for the file as detailed above. If you are running a parallel database system on several machines, your function implementation must be accessible from all the machines in your system.

The script that compiles and links your external function (such as bldvaudf.bat) should copy the resulting executable file to the default directory (sqllib/function) on the local machine, otherwise you need to explicitly copy the file yourself.

After the executable file has been copied it must be made executable. Under AIX / UNIX the hgudf file can be made executable by the following command:-

chmod a+x hgudf

But this is also done automatically by compilehgudf.sh

The process that executes the external function will not be running under your userid, but under a dummy userid supplied by the database system.

Invocation

The following simple example shows how the Htheor function might be used in a query in the development environment:-

Go into DB2 and connect to the relevant database, then enter:-

SELECT Htheor(p1, p2, p3, p4) from 'file address' where "Parameter" = 'xyzabc'

Note:- The returned SQLCODE -440 (SQLSTATE 42884) indicates that the system was unable to find any applicable function for one of your function calls. If you receive this return code, check your function path and make sure it includes the schema containing the desired function. Next, check the arguments of the function to make sure that their datatypes match to datatypes of the function parameters.

Address Space

An external function can operate in the same address space as DB2 or in it's own address space. The FENCED option specifies that your function will always be run in an address space that is separate from the database, This option entails a processing penalty due to process switching when the function is called, but it protects the integrity of the database against accidental or malicious damage that might be caused by the function. The FENCED clause is optional and is the default. After the function is thoroughly tested you may drop and re-create the function with the NOT FENCED parameter to gain a speed advantage. The NOT FENCED function runs in the same address space as the database and can damage the integrity of your data. In order to create an unfenced function you must possess SYSADM or DBADM authority.

Tables and Files

Tables

SYSIBM.SYSFUNCTIONS - The table in which the function is created.

SYSCAT.FUNCTIONS - A read only view is provided of the function.

SYSSTAT.FUNCTIONS - An updatable view of the function. Each
                     user can 'see' only those columns that they
                     are allowed to update.

32.0   Installing and Configuring DB2 UDB EEE

To install and configure DB2 UDB EEE perform the following tasks:-

Installing / Configuring DB2 UDB EEE

1.   Make the file system for the DB2 instance's home directory (home/x3an01) available across NFS to all the SP nodes.

2.   Install DB2 UDB EEE.

3.   Add a Group and a User for the DB2 Instance.

4.   Add entries to /etc/services on all SP nodes.

5.   Create the DB2 Instance.

6.   Edit .profile

7.   Edit db2nodes.cfg file.

8.   Edit the .rhosts file.

9.   Set up db2diag.log and syslog

10.   Start the DB2 Instance.

11.   Find the process id (pid) of the syslogd process.

12.   Start the instance.

13.   Create the Database, Node Groups, Table Spaces and Tables

1. Make the file system for the DB2 instance's home directory:-

(home/SP Node name) across NFS to all SP Nodes.

Smitty nfs

➔ Network File sytem (NFS)
-➔ Add a directory to the Exports List

PATHNAME of directory to export   /home/
x3an01
HOSTS allowed root access
 x3an05, x3an09, x3an13

2. Install DB2 UDB EEE

Log in as root on the first SP node xxxxxxxx
Change directory into the directory where the DB2 images
are located.

Run smitty installp then specify .(dot) as the input
directory.
Select F4 against Software to install.
Choose License for DB2 UDB EEE
Edit smit.log to find installp command that was run.

3. Add a Group and a User for the DB2 Instance

Smitty group
➔ Add a Group

Group Name db2asgrp

        To add a User:-

        Smitty User
         ➔ Add a User

          User Name....................... db2inst1
          Primary Group..................... db2asgrp
          Home Directory .... /home/x3an01/dn2inst1

Before distributing this user to the SP nodes we need
to:-
  Set it's password because to logon as a new user a
  new password must exist.
  Change it's password because by default when we first
  log in as this new user.
  The system will prompt us to change the password.

4. Adding Services Entries

Before creating the DB2 UBD EEE instance entries must
be made in the /etc/services/ file on all SP nodes to
be included in the instance. If we are using 4 database
partitions (DPs)per SP node, we need to reserve 4 ports
per SP node. To support HACMP failover, another 4 ports
must be reserved making a total of 8.

Add these lines to /etc/services:

```
DB2_db2inst1                          3000/tcp
DB2_db2inst_END                       3007/tcp
```

This means that ports 3000 to 3007 inclusive are reserved
for DB2

5. Now we are ready to create the instance:-

Log in as root at x3an01

```
x3an01[/]> cd /usr/lpp/db2-07_00/instance
x3an01[/usr/lpp/db2-07_00/instance]> ./db2icrt -ud2inst1
db2inst1
DBI1070 Program db2icrtg completed successfully
```

6. Add Profile Entries

Entries need to be added to .profile of db2inst so that
environmental variables and DB2 profile variables are
set. Add these lines to:-
/home/x3an01/db2inst1/.profile:-

```
. ~/sqllib/db2profile
# Set the default DB
db2set -I db2inst1 db2dbdft=tcpd30
```

7. Edit the db2nodes.cfg file

The file db2nodes.cfg in the directory $INSTHOME/sqllib
defines which database partition servers will be started
when a db2start is issued.

        Using vi add these lines:-

        1 x3an01 0 x3sn01
        2 x3an01 1 x3sn01
        3 x3an01 2 x3sn01
        4 x3an01 3 x3sn01
        5 x3an05 0 x3sn01
        6 x3an05 1 x3sn05
        7 x3an05 2 x3sn05
        8 x3an05 3 x3sn05

Start numbering the (database partitions —nodenums column one) at 1 so they would be synchronized with the host names of the SP nodes.

Hostname (2nd column) is defined as the first Ethernet interface (en0) on each SP node.

Logical port (3rd column) must be 0, 1, 2, 3 for the four DPs on each SP node

Switchname (4th. column) is the network interface of the switch on each SP node.

8.   Creating .rhosts entries

    Before issuing db2start we have to make sure that remote execution permission is defined across the SP nodes. In other words, if we execute 'rsh x3an05 date' from x3an01 we will not be prompted for a password. This needs to be true for all the network interfaces defined in the db2nodes.cfg file.

    Set permission by adding this line to .rhosts in /home/x3an01/db2inst1 on x3an001:-

        + db2inst1

    Where db2inst1 is the name of the instance.

Because /home/x3an01 is a file system that is shared across the SP nodes, we only have to create these entries once.

Test that the remote execute permission has been enabled for
each network interface before executing the db2start command.

9. <u>Set up the db2diag.log file</u>

   Most DB2 informational and error messages are written to
   db2daig.log. This file is by default stored in the
   $INSTHOME/sqllib/db2dump directory, which is NFS shared
   to all the SP nodes. It is a good idea to change this
   path so that the db2diag.log is written locally to each
   SP node.

   To do this, logged in as db2inst1 at x3an01, enter:-

   db2 update dbm cfg using diagpath /tmp/db2

Where /tmp/db2 is the file system that exists locally on each
SP node. Since this parameter (diagpath) is defined at the
database manager level this command is executed once for the
DB2 instance.

All the DPs on the same SP node will write to the same file,
so output from DPs 1,2,3 and 4 will go to /tmp/db2/db2diag.
log on x3an01.

10. <u>Setting up syslog.conf</u>

     Some DB2 informational and error messages apply to
     the SP node and are Independent of the DB2 instance.
     These messages are captured by configuring the AIX
     SYSLOG daemon:-

     Add to /etc/syslog.conf:- user.warn
                                 /tmp/db2/syslog.db2

   Create the /tmp/db2/syslog.db2 file:-
                 touch tmp/db2/syslog.db2

11. Find the process id (pid) of the syslogd process:-

    ps -ef|grep syslogd

    Send a -1 signal to this process using Kill:-

        Kill -1 <syslogd-pid>

        Run this set of commands on each SP node.

12. Start the instance:-

    Now we are ready to start the instance:-

    db2start

    You can start a single DP server:- db2start nodenum 16

13. Create the Database, Nodegroups, Table Spaces and Tables:-

    Db2 terminate

33.0 Parallelism

One of the most important concepts of UDB, is UDB parallelism and the concept of a partition.

In a parallel system each database can be split into several parts called partitions. Each table in the database can have some of its rows in each partition. It is helpful to visualize each partition as running on a separate machine, although it is possible for more that one partition to be assigned to the same machine.

Each database partition has it's own log and it's own set of indexes.

Intra-partition parallelism refers to simultaneous processes within a single partition.

Inter-partition parallelism refers to simultaneous processes in multiple partitions. These two kinds of parallelism are independent of each other, and their relative importance depends on your hardware configuration.

Intra-Partition Parallelism

IntraPP is often used on a symmetric multiprocessor (SMP) machine, in which multiple processors share common memory and disks, To exploit IntraPP, the optimizer generates access plans that contain multiple threads that can be active simultaneously during processing of an SQL statement.

Since intraPP takes place within a single database partition, all the threads have access to all the data in the partition.

The number of threads in an access plan, called the "degree" of the plan, may be more or less than the number of physical processors on the system.

To use IntraPP with degree 4, the optimizer may generate a plan with four threads each scans a different part of the table.

Even if four threads are shared by two or three processors, a significant performance improvement can result.

IntraPP is completely transparent to SQL it places no limitations on SQL statements.

To exploit intraPP set the database manager configuration parameter INTRA_PARALLEL to YES. The optimizer will then generate multiple threads and the run time system will execute these plans using multiple processors.

BIND with DEGREE ANY - no limit on the degree chosen by the optimizer. Default is when configuration DFT_DEGREE is 1. CURRENT_DEGREE can be set to ANY default is 1 as per the DFT_ DEGREE parameter.

MAX_QUERYDEGREE and SET RUNTIME DEGREE also limit the number of threads running in a plan.

34.0   Unix Data Splitters

The data must be split into several sub sets based on the value
of a partitioning key, then each subset must be loaded into
it's respective partition. UDB provides two utilities called
Splitter and Autoloader.

Splitter is provided in the form of a program named db2split
found in sqllib/bin, that runs on all UDB platforms. The
Splitter's job is to scan an input file, analyze the data
according to a specified partitioning key, and prepare the data
for loading into a node group in a partitioned database. The
input file can be in ASC or DEL (delimited ASCII) format or BIN
(binary numeric) or PACK format (packed decimal).

The splitter is controlled by a configuration file that specifies
the names of the input and output data files, the partitioning
key, the node numbers of the partitions in the nodegroup, and
various other control parameters. The configuration file specifies
one of the two following modes:-

ANALYZE mode:- The Splitter scans the input file, analyzes its
data distribution in the partitioning key columns, and generates
an optimal partitioning map that can later be used to distribute
data uniformly among the partitions in the nodegroup. Before
loading the data, you can install the generated partitioning
map in the nodegroup to be loaded by
means of a REDISTRIBUTE NODEGROUP command.

PARTITION mode:- In this mode the Splitter scans the input file
and splits it into several load files, each containing data
ready to be loaded into one of the partitions of a nodegroup.
The data is split by hashing the partitioning key of each row
into one of 4096 hash buckets, then mapping the hash buckets
into partitions using a partitioning map.

The partitioning map is specified in the configuration file and
may come from any of the following sources:
   1. It might have been generated by a previous run of the
      Splitter in ANALYZE mode.It might be a default partitioning
      map that distributes the hash buckets uniformly among the
      partitions.
   2. It might be the partitioning map that is currently is
      use in the nodegroup where the data will be loaded. This

partitioning map can be retrieved from the System Catalog tables named NODEGROUPS and PARTITIONMAPS or by using a utility called db2gpmap.

3. It might have been constructed manually. A partitioning map is simply an array of 4096 two-byte integers that indicate node numbers associated with the 4096 hash buckets.

The load files generated by the Splitter have name suffixes that identify the partitions where the data is to be loaded e.g. LOADDATA.000, LOADDATA.001, LOADDATA.002, LOADDATA.003. Each load file contains a header that records the partitioning map that was used to generate it. When loading data from these files the LOAD utility generates an error message if the current partitioning map that is being loaded is not the same as the partitioning map by which the load files were generated.
To invoke splitter:- db2split -c db2split.cfg

35.0   UNIX Autoloader

The Autoloader is provided in the form of a program named db2autold, found in the sqllib/misc library, that runs only on the UDB Enterprise-Extended Edition (now ESE). It provides an interface with the same functionality as the Splitter with some additional features. It can transfer it's input data from a remote system via FTP and after splitting the data it can load it into a selected nodegroup in a partitioned tablespace.

Like the Splitter, the Autoloader is controlled by a configuration file that specifies the input, the output and the type of operation desired.

The Autoloader can be invoked in the following type of roles:-

1. ANALYZE role:- It scans an input dataset and generates the distribution of data among the partitions in the selected nodegroup.

2. SPLIT_ONLY role:- This role is similar to the PARTITION role of the Splitter. Using a given partitioning map, it splits an input data set into multiple load files ready for loading into the partitions of a nodegroup.

3. LOAD_ONLY role:- This role actually loads the files generated by the SPLIT_ONLY mode into the partitions of a nodegroup by invoking the Load utility on each of the partitions, using a LOAD command that is specified in the configuration file.

3. SPLIT_AND_LOAD role:- This role combines the functionality of the SPLIT_ONLY and LOAD_ONLY roles. In this role the splitter output is piped directly to the load utilities running on the various machines in the nodegroup. As a result, the split data does not need to be stored in intermediate files, and the total time for splitting and loading is minimal.

A sample Autoloader configuration file named autoloader.cfg can be found in the sqllib/samples/autoloader directory. You will need to edit this file before use, following the comments provided in the file to customize it for your input and output. The following command invokes the Autoloader under the control of the sample configuration file:-

```
db2autold  -c  autoloader.cfg
```

36.0   DB2 LOAD (a.k.a. High Speed Bulk Loader)

The Load facility to use is the DB2 High Speed (HS) Bulk Loader otherwise known as the DB2 Load.

Loads are done using AIX and not the DB2 UDB Control Center or Command Center. The script to execute all Loads will be executed from a Unix shell, having a name and suffix as follows:-

Name:- loadxxxxxxxxxxx   where xxxxxxxxxxxx is the name of the table(s) to be loaded.

Period:-   .

Suffix:-   sh   - Lowercase sh

So a load ksh file to load Claims would look like this:- loadclms.sh

The sequence necessary to execute a load ksh is as follows:-

1. at now (this executes the load in batch mode).

2. loadtable.sh   > xxxxxxxx yyyyyyyy loadzzzzzoutput.txt

3. ctrl   D

In the above sequence:- 1 ensures the load is started immediately in batch mode.

> 2 loadtable.sh is the load .ksh file which identifies the file containing the Load Instructions.

> > Says that the output is going to be piped to an output file.

> xxxxxxxxx the name of the parent or single file to be loaded

> yyyyyyyyy the name of the child file to be loaded if present

> 3. ctrl d - kicks off the Load

A few Load Performance Parameters

The IBM DB2 HS (High Speed) Bulk Loader will automatically switch RI off when loading for the sake of speed.

The REPLACE verb is faster than INSERT because REPLACE is done in buffered blocks rather than one at a time as in the case of INSERT.

FASTPARSE reduces DB2 parsing thus increasing speed.

ANYORDER tells DB2 not to bother with checking for a particular sequence, it loads what it gets.

DISK_PARALLELISM when set is based on the number of containers associated with the table spaces loaded (Configuration File).

DATA_BUFFER - The buffer size is allocated directly from the utility heap (Configuration File).

SORT BUFFER - Set this to as large a value as possible for the larger tables (Configuration File).

INTRA_PARALLEL - Make this YES in the Database Manager Configuration file (Configuration File).

Exception Table - All load errors will be written as Load Messages and also written to the Exception Table. You may create an Exception Table like this:-

CREATE MY_EXCEPTION_TABLE LIKE MY_BASE_TABLE
ALTER TABLE MY_EXCEPTION_TBL ADD COLUMN TS TIMESTAMP ADD COLUMN MSG CLOB(32K);

Although the Load switches off RI, it "keeps an eye out" for a child table being loaded and when it detects a child it immediately puts the child table into "Check Pending" state. Which means that the child will be loaded but will not be able to be used until the Integrity switch is set (see below).

When loading, the Load supplies values for the Identity Column of the tables as per the specifications in the DDL (Start Number and increment by value).

After a Parent / Child combination are loaded, then the Child's

Identity Column needs to be made the same as the Parent's Identity Column as well as the natural key(s). How is this done?

The natural key of the parent table is used as the index to scan the child table. When the natural key of the parent equals the natural key of the child, then the value of the Identity Column in the parent is put in the predetermined Identity Column slot of the child table, thus ensuring RI for both the natural keys of parent and child, and the Identity Columns of parent and child are ready for Referential Integrity.

Example of a SQL Script follows:-

```
P_TABLE              = Name of the Parent Table.
C_TABLE              = Name of the Child Table.
NAT_KEY_P_1          = First Natural Key of the Parent Table.
NAT_KEY_P_2          = Second Natural Key of the Parent Table.
NAT_KEY_C_1          = First Natural Key of the Child Table.
NAT_KEY_C_2          = Second Natural Key of the Child table.
P_ID_COL             = Name of the Parent Table's Identity Column.
C_ID_COL             = Name of the Child Table's Identity Column
- predetermined slot.

UPDATE C_TABLE SET CT.C_ID_COL = PT.P_ID_COL
SELECT NAT_KEY_P_1, NAT_KEY_P_2 FROM P_TABLE AS PT
NAT_KEY_C_1, NAT_KEY_C_2 FROM C_TABLE AS CT
WHERE   PT.NAT_KEY_P_1 = CT.NAT_KEY_C_1
AND       PT.NAT_KEY_P_2  = CT.NAT_KEY_C_2;
```

Set Referential Integrity (RI)

After a parent table is loaded, followed by a child table, now is the point at which RI is applied. This is done by the set integrity statement:-

```
SET INTEGRITY FOR TABLE P_TABLE IMMEDIATE CHECKED FORCE GENERATED;

SET INTEGRITY FOR TABLE C_TABLE IMMEDIATE CHECKED FORCE GENERATED;
```

Load script

So the loadtable.sh .ksh file for loading a parent and child
table combination would look like this:

```
#! /bin/ksh
date
db2 connect to hsddb01 user hsdprd01 using ibmdb2;
db2 set current schema = HSDMPROD;

date


db2 "LOAD FROM P_TABLE_FLAT_FILE OF DEL MODIFIED BY ANYORDER
FASTPARSE SAVECOUNT 50000 REPLACE INTO P_TABLE INDEXING MODE
REBUILD FOR EXCEPTION P_TABLE_EXCEP";

db2 "LOAD FROM C_TABLE_FLAT_FILE OF DEL MODIFIED BY ANYORDER
FASTPARSE SAVECOUNT 50000 REPLACE INTO C_TABLE INDEXING MODE
REBUILD FOR EXCEPTION C_TABLE_EXCEP ";

db2 UPDATE C_TABLE SET CT.C_ID_COL = PT.P_ID_COL
    SELECT NAT_KEY_P_1, NAT_KEY_P_2 FROM P_TABLE AS PT
    NAT_KEY_C_1, NAT_KEY_C_2 FROM C_TABLE AS CT
    WHERE  PT.NAT_KEY_P_1 = CT.NAT_KEY_C_1
    AND PT.NAT_KEY_P_2  = CT.NAT_KEY_C_2;

db2 SET INTEGRITY FOR TABLE P_TABLE IMMEDIATE CHECKED FORCE
GENERATED;
db2 SET INTEGRITY FOR TABLE C_TABLE IMMEDIATE CHECKED FORCE
GENERATED;
date
db2 terminate;
```

Note:- The SAVECOUNT 50000 parameter does a COMMIT every
        50000 rows loaded.

37.0 DB2 Import

The IMPORT utility inserts data from an input file into a table
or view. You can either replace or append data if the table or
view already contains data.

If the existing table contains a primary key or unique constraint
that is referenced by a foreign key in another table, data
cannot be replaced, only appended.

With the IMPORT utility, you can specify how to add or replace
the data into the target table. You must be connected to the
database to use the utility. If you want to import data into
a new table using the CREATE option, you must have SYSADM or
DBADM authorities or CREATEDB privilege for the database. To
replace data in a table or view you must have SYSADM or DBADM
authorities or CONTROL privilege for the table or view. If you
want to add data to an existing table or view, you must have
SELECT and INSERT privileges for the table or view.

The complete syntax of the IMPORT command follows:-

```
IMPORT  FROM  filename OF  { IXF  |  ASC |  DEL  |  WSF }
[LOBS FROM lob-path  [  {, lob-path } ...... ]  ]
[MODIFIED BY filetype-mod.... ]
[METHOD  { L (col-start  col-end  [  {, col-start  col-end}
.....] )
[NULL INDICATORS  (col-position [  { , col-position}..... ]  )
]  |
N ( col-name [  {, col-name} ...... ]  )  |
P ( col-position  [ , col-position} ..... ]  ) } ]
[COMMITCOUNT n]  [RESTARTCOUNT n]  [MESSAGES message-file ]
{{ INSERT  |  INSERT_UPDATE  |  REPLACE  |  REPALCE_CREATE}
INTO  table-name [ ( insert-column, .......) ]
|  CREATE INTO table-name  [ ( insert-column , ..... ) ]
IN tablespace-name  [ INDEX IN tablespace-name ]
LONG IN tablespace-name ] ] }
```

Many of the parameters of the IMPORT command have the same
definition as those of the EXPORT command. The parameter filetype
is the same. The IMPORT command supports the same file types
as EXPORT, and the plain ASCII file type ASC. However, some of
the parameters are different:

filetype-mod – this information is unique to the file format (DEL, WSF, or IXF) that was specified in the filetype parameter. If you specify LOBSFILE within the filetype-mod string, ASC, DEL or IXF input files contain the names of the files that have LOB data in the LOB column.

For the ASC file format, filetype-mod can be used to indicate that the trailing blanks after the last non-blank character are to be dropped when importing into a variable-length database field. If you specify a T within the filetype-mod string, this indicates that trailing blank spaces are to be truncated. If the filetype-mod parameter is not specified within the command string, blank spaces are kept.

For DEL (delimited ASCII) file format, filetype-mod selects characters to override column delimiters, character string delimiters, decimal point characters and positioning of decimal points.

For IXF, you can suspend the comparison of the code page values in the input file with the application and the database. There are two options to consider. The FORCEIN option tells the IMPORT utility to accept the data even if the codepages do not match, and perform no transportation on them. The INDEXIXF option tells IMPORT to drop all the indexes that are currently defined on the existing table. New ones will be created using the index definition found in the PC/IXF file. You can only use this option when replacing existing data. You cannot use this option with a view or when insert-column is specified.

For WSF, filetype-mod is ignored.

METHOD L – The METHOD parameter with the L option specifies the start and end column number of the data being imported. This option must be used for ASC files.

METHOD N – This parameter specifies the names of the columns to be imported.

METHOD P – The METHOD parameter with the P option specifies the order of column numbers to be imported. If no method is selected, then the default columns are used during the import.

COMMITCOUNT n – A commit will be done every n records.

RESTARTCOUNT n – This specifies that an import will be started at record n + 1. The first n records are skipped.

INSERT – Adds the imported data to the table without changing the existing table data.

INSERT_UPDATE – Adds rows of imported data to the target table or updates existing rows of the target table with matching primary keys.

REPLACE – This option deletes all existing data in the table and inserts the imported data. The table definition and index definition are not changed. You can use REPLACE only on an existing table.

CREATE – This option creates the table definition and row contents. If the data is exported from another DB2 UDB database, indexes are also created. This option can only be used with IXF files. You can specify the table spaces for normal, LOB and index data.

REPLACE_CREATE – If the table exists, all the data in that table is deleted and the imported data is inserted without changing the table and index definitions. If the table does not exist, the table definitions are created and data is inserted. This can only be used with IXF files. Also, if data is exported from a DB2 UDB database, indexes are also created.

Note:- When you right-click on a table in the Control Center and click on IMPORT, you can specify the source file for import and select options related to the file type, large objects, and column positioning.

IMPORT Considerations

The following information is required when importing data to a table or view:-

The path and the input file name where the data is stored.

The name or alias of the table or view where the data is to be imported.

The format of the data in the input file. This format can be IXF, DEL, ASC or WSF.

Whether the data in the input file is to be inserted, updated, replaced or appended to the existing data in the table or view.

You may also provide:-
1. A message file name. 2. The number of rows to insert before committing changes to the table. 3. The number of records in the input file to skip before beginning the import. 4. The names of the columns within the table or view in which the data is inserted.

The IMPORT utility will issue a COMMIT or a ROLLBACK statement, depending on it's success.

A summary of file types supported by Import, Export and Load follows:-

Function	Delimited ASCII (DEL)	Nondelimited ASCII (ASC)	Integrated Exchange Format (IXF)	Worksheet File (WSF)
EXPORT	YES	NO	YES	YES
IMPORT	YES	YES	YES	YES
LOAD	YES	YES	YES	NO

Note:- Delimited ASCII **(DEL)** files consist of streams of data values, ordered by row and by column within each row. Values are separated by column delimiters (default is a comma) and rows are separated by row delimiters (default is a new line character).

Nondelimited ASCII **(ASC)** files are similar to delimited ASCII files, except that the columns of data within each row are found in fixed positions and therefore do not need to be marked by delimiters. When using Nondelimited ASCII files the Import and Load utilities must specify the exact format of the file (i.e. the character positions assigned to each column of data).

Integrated Exchange Format **(IXF)** files are the preferred format for transferring data between databases managed by UDB on the same platform or on different platforms. For example, an IXF file can be used to move data between a database running under Windows NT and a database running under AIX, automatically compensating for the different numeric formats used on these two platforms.

Worksheet **(WSF)** files are intended for interchange of data between UDB and certain versions of Lotus 1-2-3 and Symphony products. WSF files can contain both column names and data.

1. Copy the flat files to be "loaded" to the database(s) to your C drive. Use FTP or other transfer software agent.

2. Format the files to DEL format (Delimited ASCII). Example:-

   "Screw", 5.10, 28, "ABC Tools", 19960211, "10.34.05"
   "Nail", 18.00, 8, "Tools ", 19980401, "09.34.05"
   "Pliers", 24.50, 10, "Bob's House of Tools", 19971230, "21.34.05"
   "Adjustable Tool ", 32.29, 15, "Tools Unlimited", 19980215, "21.34.05"

DELimited ASCII Rules

Delimited ASCII (type DEL) files consist of streams of data values, ordered by row and by column within each row. Values are separated by column delimiters (by default a comma), and rows are separated by row delimiters (by default a new line character in UNIX and <u>a carriage return/linefeed sequence in Windows – Hit Enter/Return at the end of each line</u>). Character string values are enclosed in string delimiters (by default a double quote). Null values are denoted by missing data (column delimiters separated by spaces or nothing → ""). When you import a delimited ASCII file you can override the default delimiters with your own delimiter characters.

After you have formatted all your file(s) on the C drive in DEL format, we will use IMPORT to insert the data to your tables. If you do not have enough space on your C drive you can still do this process one file at a time.

38.0   DB2 Export

The **EXPORT** utility is used to export data from a database to a file. Data can be exported in several different file formats that can be used with other programs. You must already be connected to the database from which data is being exported, and you must have SYSADM or DBADMIN Authority; or CONTROL or SELECT privilege.

The full syntax of the Export command is as follows:-

```
EXPORT TO filename OF {IXF | DEL | WSF}
[LOBS TO lobs-path [ {,lob-path} …..]   ]
[LOBFILE lob-file [ {, lob-file} ……..]
[MODIFIED BY {filetype-mod …... }]
[METHOD N (column-name [   {, column-name} ……]   )   ]
[MESSAGES message-file]
select-statement
```

Some of the parameters of the EXPORT command are as follows:-

filename – Specifies the name of the file to which data is exported. If the path is omitted, the current working directory is used.

filetype – Specifies the format of the data in the output file.

DEL (Delimited ASCII format) is the same format you can use with the LOAD utility. DEL makes it possible to exchange a wide variety of files between different database managers and file manager programs.

WSF format allows you to create data that can be used by spreadsheet applications.

IXF (Integrated Exchange Format, PC version) makes it possible to import into the same or another type of database manager table. The definition of the table as well as any existing indexes are saved in the IXF file header.

filetype-mod – Specifies additional information unique to DEL or WSF file formats that was specified in the filetype parameter. The parameter is ignored with IXF file types.

METHOD N - This parameter indicates that the specified column names are to be used for the output file. if this parameter is not specified, the default names (column names of the existing table) are used for the columns. This parameter is only valid for IXF and WSF files.

msgfile - Specifies the destination of warning and error messages that occur during the export. If msgfile is omitted, the messages that occur are written to the screen. If you do not specify the complete path to the file, EXPORT uses the current directory as the destination. If you specify the name of a file that already exists, EXPORT will overwrite the file.

SELECT statement - This specifies the select statement that will be used to extract the records that will be exported.

In DB2 UDB, by right clicking on a table in the Control Center and clicking on Export, you can export data from the table through the graphical export interface. You can also specify file names for LOBs in the table, and specify explicit column names for the data exported to IXF or WSF format.

Export Example follows:-

```
EXPORT TO myfile.ixf OF IXF MESSAGES msgs.txt
SELECT * FROM db2test.table;
```

39.0 AGENTS AND THREADS

Client Programs

Client programs run remotely or on the same machine as the database server. They make their first contact with the database through a listener. A coordinator agent (db2agent) is then assigned to them.

Listeners

Client programs make initial contact with communication listeners, which are started when DB2 is started. There is a listener for each configured communication protocol, and an inter-process communications (IPC) listener (db2ipccm) for local client programs. Listeners include:

- **db2ipccm**, for local client connections

- **db2tcpcm**, for TCP/IP connections

- **db2snacm**, for APPC connections

- **db2tcpdm**, for TCP/IP discovery tool requests

To find out if there was a problem activating a listener, consult the db2diag.log at the server. Whether they are local or remote, they are allocated a corresponding coordinator agent (**db2agent**). When the coordinator agent is created, it performs all database requests on behalf of the application.

In some environments in which the *intra_parallel* database manager configuration parameter is enabled, the coordinator agent distributes the database requests to subagents (**db2agntp**). These agents perform the requests for the application. Once the coordinator agent is created, it handles all database requests on behalf of its application by coordinating subagents (**db2agntp**) that perform requests on the database.

A coordinator agent may be:

- Connected to the database with an alias. For example, "db2agent (DATA1)" is connected to the database alias "DATA1".

- Attached to an instance. For example, "db2agent (user1)" is attached to the instance "user1".

Idle agents reside in an agent pool. These agents are available for requests from coordinator agents operating on behalf of client programs, or from subagents operating on behalf of existing coordinator agents. The number of available agents is dependent on the database manager configuration parameters *maxagents* and *num_poolagents*.

db2udfp

Fenced user-defined functions (UDFs) run outside of the firewall.

db2dari

Fenced stored procedures run outside of the firewall.

Database Threads/Processes

The following list includes some of the important threads/processes used by each database:

- **db2pfchr**, for input and output (I/O) prefetching

- **db2pclnr**, for buffer pool page cleaners

- **db2loggr**, for manipulating log files to handle transaction processing and recovery

- **db2dlock**, for deadlock detection.

Database Server Threads/Processes

The system controller (**db2sysc**) must exist in order for the database server to function. Also, the following threads/processes may be started to carry out various tasks:

- **db2resyn**, the resync agent that scans the global resync list

- **db2gds**, the global daemon spawner on UNIX-based systems that starts new processes

- **db2wdog**, the watchdog on UNIX-based systems that handles abnormal terminations

- **db2fcmdm**, the fast communications manager daemon for handling internodal communication (used only in DB2 Enterprise - Extended Edition)

- **db2pdbc**, the parallel system controller, which handles parallel requests from remote nodes (used only in DB2 Enterprise - Extended Edition)

- **db2panic**, the panic agent, which handles urgent requests after agent limits have been reached at a particular node (used only in DB2 Enterprise - Extended Edition)

Differences between Intel and UNIX

The Intel systems supported by DB2 (OS/2 and Windows) differ from UNIX-based environments in that the database engine is multi-threaded, not multi-processed. In the Intel systems, each of the dispatchable units on the agent side of the firewall is a thread under the process `db2sysc`, allowing the database engine to let the operating system perform task-switching at the thread level and not the process level. For each database being accessed, there are other threads started to deal with database tasks (for example, prefetching).

Another difference is in the handling of abends. There is no need for a "watchdog" process in Intel systems, because these systems ensure that the allocated resources are cleaned up after an abnormal termination. Thus, there is no equivalent of the **db2wdog** process on the Intel systems. In addition, the **db2gds** process/thread is not needed on the Intel systems, which have their own mechanisms for starting threads.

40.0   Compression

Using the COMPRESS clause in the CREATE TABLESPACE and ALTER TABLESPACE SQL statements allows you to compress data in a table space or in a partitioned table space. In many cases, using the COMPRESS clause can significantly reduce the amount of DASD space needed to store data, but the compression ratio you achieve depends on the characteristics of your data. In addition, with compressed data, the amount of I/O required to scan a table space can be reduced.

Performance Considerations

To estimate how well a given table space will compress you can run the standalone utility N1COMP, however, consider the following when making your decision:-

Compressing data can result in higher processor cost, depending on the SQL work load. However if you use IBM's synchronous data compression hardware processor time is significantly less than if you use just the DB2-provided software simulation or an edit or field procedure to compress the data. Indexes are not compressed.

The processor cost to decode a row using the COMPRESS clause is significantly less than the cost to encode that same row. This rule applies regardless of whether the compression uses the synchronous data compression hardware or the software simulation built into DB2.

When rows are accessed sequentially, fewer I/Os might be required to access data stored in a compressed table space. However, there is a trade off between reduced I/O resource consumption and the extra processor cost for decoding the data.

If random I/O is necessary to access the data, the number of I/Os will not decrease significantly, unless the associated buffer pool is larger than the table and the other applications require little concurrent buffer pool usage.

Depending on the SQL work load, as the degree of compression increases it can lead to:-

High buffer pool hit ratios, Fewer I/Os, Few "get page" operations.

The data access path DB2 uses affects the processor costs for data compression.

In general, the relative overhead of compression is higher for table space scans and less costly for index access, consequently, doing Explains on the relevant SQL and Tables should point you in the right direction before a "let's compress" decision is made.

Also, some types of data are better than others for compression. Data with hexadecimal content are not good for compression, and longer rows (row length above 208 characters) are better candidates for compression than shorter rows. For each row compressed there is an 8 byte DB2 overhead attached. Average compression gains measured by IBM over an extended period of time with various data combinations average out at 61% space saved.

Connections

There is no need for the connection to be re-established again and again whenever it is needed to be used. Once a connection has been created and placed n a pool, an application can reuse that connection without performing a complete connection process. The connection is pooled when the application disconnects from the ODBC data source, and will, later, be given a new connection whose attributes are the same as the connection the application previously relinquished.

ODBC connections in MTS (Microsoft Transaction Server) COM objects have connection pooling turned on automatically whether or not the COM object is transactional. For multiple MTS COM objects participation in the same transaction the connection can be re-used between two or more COM objects in the following manner
which may cause a problem:-

Let us say there are two COM objects COM1 and COM2 that connect to the same ODBC data source and participate in the same transaction. After COM1 connects and completes it's work it then disconnects and the connection joins the connection pool. However this connection is reserved for use by other COM objects of the same transaction. It will be available to other transactions only after the current transaction ends.

When COM2 is invoked in the same transaction, it is given the recently pooled connection released by COM1. MTS ensures that the connection can only be given to COM objects participating in the same transaction.

On the other hand, if COM1 does not disconnect, it will tie up the connection until the transaction ends. When COM2 is invoked, in the same transaction, a separate connection will be acquired. Subsequently, this transaction ties up two connections instead of one.

This will result in timeouts, deadlocks and wait states. It could also affect WebSphere applications.

This reuse of the connection feature for COM objects participating in the same transaction is preferable as follows:-

- It uses fewer resources in both the client and the server. Only one connection is needed.
- It eliminates the possibility that two connections participating in the same transaction (accessing the same database server and accessing the same data) can lock one another, because DB2 servers treat different connections from MTS COM objects as separate transactions.

- Alternatively, the environment can be patched using an IBM Fix Pack. There are two fix packs from IBM available. Fix Pack 4 & 5 and Fix Pack 6. Fix Pack 6 contains the eFix which is the preferred method.

41.0 Constraints On and Off

The syntax used to turn off constraints on tables is:-

SET INTEGRITY FOR <table name> OFF

This will turn the integrity checking for one or more tables off. This includes the following:-

- Check constraint checking
- Referential constraint checking
- Datalink integrity checking

The generation of values for generated columns is also turned off. Moreover, the immediate refreshing of data is turned off for summary tables that are defined REFRESH IMMEDIATE.

Tables that have integrity checking turned off are placed in a check pending state. Primary key and unique constraints continue to be checked.

One version of the syntax used to check for constraints on tables is the following:-

SET INTEGRITY FOR <table name> ALL IMMEDIATE UNCHECKED

The ALL clause in the statement specifies that all integrity options are to be turned on. You have the choice of turning them all on in this way, or you can list each integrity option that you wish to have enforced. One or more can be selected in a single statement.

The table in this example is not checked for integrity violations. If the table is a summary table, immediate refreshing is turned on.

If the ALL clause is not used, one or more integrity options can be chosen for each table. These choices include FOREIGN KEY, CHECK, DATALINK RECONCILE PENDING, SUMMARY and GENERATED COLUMN.

The table is left in check pending state when at least one of the five possible integrity options is left off for that table. If the parent of a dependent table is in check pending state, the foreign key constraints of the dependent table cannot be marked to bypass checking, but check constraints checking can be bypassed.

Here is an example of the use of the SET INTEGRITY statement:-

SET INTEGRITY FOR EMP_ACCT OFF;
ALTER TABLE EMP_ACCT ADD CHECK (EMPSTDATE <= EMPENDDATE);
ALTER TABLE EMP_ACCT ADD FOREIGN KEY (EMPNO) REFERENCES
EMPLOYEE; SET INTEGRITY FOR EMP_ACCT IMMEDAITE CHECKED;

42.0 Speeding up Performance

One – Create a temporary tablespace called TEMPSPACE2 using one SMS container on a different operating system file located on a different physical disk. You should cut elapsed time by around 25% for the query.

Two – Add tempspace containers for parallel I/O. Place these containers on disks that are not used by the database. If the configuration has other short comings, then the results of this technique may not be as dramatic as expected, but in the worst case expect to see an improvement of 15% for the query.

Three – Your EXTENTSIZE and PREFETCHSIZE will probably be at the 32 4K pages mark. It really does not matter. DB2 has the ability to prefetch data asynchronously in anticipation of the CPU's ability to process it, thus minimizing the impact of I/O service delays. PREFETCH dictates the quantity of data to asynchronously retrieve for each prefetch request. Generally, PREFETCHSIZE should be a multiple of EXTENTSIZE, and the multiple should be equal to the number of containers. For performance set the PREFETCHSIZE to 128 (example:- four containers multiplied by EXTENTSIZE 32). This should get you from about 5% to 10% improvement in performance.

Four – Use parallelism, by updating the database configuration parameter using INTRAPARALLEL YES (db2 update dbm cfg using INTRAPARALEL YES) and set the default degree of parallelism to ANY (db2 update db cfg for DBNAME using DFT_DEGREE ANY). While this technique will improve elapsed time it could be expensive in CPU time, unless you have ample memory. If you have sufficient memory this technique should help reduce query time by about 4%.

Five – Increase the size of the IBMDEFAULT bufferpool by a factor of 16, so if the IBMDEFAULT bufferpool is 4 MB, make it 64 MB. This will have a major impact on query time and should reduce the average query time by 30%.

Six – Move from SMS to DMS and watch the query time drop by at least 5%. Thus each large table (over 1 million rows) should be residing in it's own DMS space.

Seven – Reduce sort overflows. When a sort is larger than the SORTHEAP size, a sort overflow occurs. The default SORTHEAP size is 256 4K pages. Increase the SORTHEAP size to 512 4K pages. This should avoid sort overflow, which is expensive.

Eight – If you use RAID devices, then before creating tablespaces on RAID devices use the db2set command to set the environment variable DB2_STRIPED_CONTAINERS=ON. By doing so you will indicate that tablespaces are defined on RAID devices. You can find the Tablespace Id by using the command db2 list tablespaces. If tablespaces 3, 4, 5, 6 and 7 are on RAID devices, the proper command to issue would be:- db2set DB2_PARALLEL_IO=3,4,5,6,7.

Put it all together by:- Using points one through eight in conjunction with each other, you will see a vast improvement in query execution times.

43.0   Java

Relational databases play a key role in most applications that need a persistent data store, but they are not the driving force behind the development and selection of various programming models, frameworks and architectures. Application development priorities are usually driven by factors outside of the scope of relational database management systems (RDBMS), such as whether J2EE or .NET is the preferred programming model for a particular shop. In recognition of the fact that you may make a commitment to one programming architecture and someone else might make a commitment to a different one, the DB2 strategy is to provide the best integration with any choice you make.

While this may be a bit of oversimplification, the current application development landscape is dominated by three distinct types of programming:
- Developing applications based on the J2EE programming model using Java.
- Developing Windows, Web, and .Net-based applications in Microsoft environments.
- Developing UNIX, Linux, and Windows applications using the C programming language.

For Java programmers, DB2 offers two application programming interfaces (APIs): JDBC and SQLJ. To enable the JDBC programming interface, DB2 Version 7.2 offers different JDBC drivers. SQLJ is an API developed by a consortium of companies and provides both a simpler programming model and an ability to create static SQL. While SQLJ usage is on the rise, the tools for SQLJ programming are not as widespread as those for JDBC.

Both vendors and in-house programmers developing native (as opposed to Web) applications for UNIX, Linux, and Windows platforms commonly use the C programming language and one of the following DB2 APIs:
- Open DataBase Connectivity (ODBC).
- DB2 CLI (Call Level Interface).
- Embedded SQL.

Embedded SQL is an interface that is very familiar to mainframe programmers, but in the client-server world both DB2 CLI and ODBC are much more widely used. DB2 CLI and ODBC are syntactically and grammatically almost identical, with DB2 CLI being a slight superset of ODBC.

DB2 has a rich support for the J2EE programming model, allowing applications developers to access DB2 data from a variety of methods.

You can access DB2 data by putting the Java program into a module in one of the following ways:
- Java stored procedures,
- Java user-defined functions (UDFs),
- Java applets, executing within a Java enabled web browser.

Because of the size of the applets and some functionality restrictions, many application designs cannot be fully implemented as applets. The DB2 Application Development client provides the full JDBC and SQLJ interfaces.

For a large-scale cluster of WebSphere Application Servers, each of the WebSphere servers could install a DB2 Runtime Client (which includes the JDBC driver).

DB2's Java support includes support for JDBC, a vendor-neutral dynamic SQL interface that provides data access to your application through standardized Java methods. JDBC is similar to DB2 CLI in that you do not have to precompile or bind a JDBC program. As a vendor-neutral standard, JDBC applications offer increased portability – a required benefit in today's heterogeneous business infrastructure. An application written using JDBC uses only dynamic SQL. The JDBC interface imposes additional processing overhead for obvious reasons.

The new trends in application development have created a myriad of naming schemes and conventions that can often prove to be the most confusing part. Let's look at the basic information you need to know to make sense of it all: versions, drivers, and kits.

A common source of confusion with Java is the fact that there are different versions of JDBC and different types of JDBC drivers that can be used by different versions of Java. What's more, the versions of Java have their own nicknames and their own Java Development Kits (JDKs).

DB2 Universal Database, for the most part, supports JDBC Versions 1.2, 2.0, and 2.1 – with some operating system restrictions. Currently, work is under way for Sun Microsystems to complete the JDBC Version 3.0 specification which has just recently moved into its final draft.

Though not an official statement from within the DB2 development labs, there has not been any changes between JDBC Version 2.0 and JDBC Version 2.1 with respect to a DB2 environment, yet! DB2 Version 7.2 provides certifiable support for different versions of JDBC. However, you can still attain JDBC-specific certification for different versions and not fully implement the standard (minor exceptions are allowed).

44.0 Java Driver Types

A JDBC Type 1 driver is built into Java and basically provides a JDBC-ODBC bridge. DB2 Version 7 does support this type of driver since it is just an ODBC conversion; however, it is typically not used any more.

The DB2JDBC Type 2 driver is quite popular and is often referred to as the **app driver**. DB2 Version 7 supports an app driver. The app driver name comes from the notion that this driver will perform a native connect through a local DB2 client to a remote database, and from its package name (COM.ibm.db2.jdbc.app.*).
In other words, you have to have a DB2 client installed on the machine where the application that is making the JDBC calls/runs. The JDBC Type 2 driver is a combination of Java and native code, and will therefore always yield better performance than a Java-only Type 3 or Type 4 implementation.

Programmers using the J2EE programming model (for example, using SQL on a Web application server) will gravitate to the Type 2 driver as it provides top performance and complete function (including the JTA interface); it is also certified for use on J2EE servers (for example, IBM WebSphere. JDBC Type 2 drivers can be used to support JDBC 1.2, JDBC 2.0, and JDBC 2.1.

The JDBC Type 3 driver is a pure Java implementation that must talk to middleware that provides a DB2 JDBC Applet Server. This driver was designed to enable Java applets for access to DB2 data sources. In this scenario, the application may talk to another machine where a DB2 client has been installed.

The JDBC Type 3 driver is often referred to as the **net driver**, appropriately named after its package name (COM.ibm.db2.jdbc.net.*). DB2 Version 7 supports the net driver. The JDBC Type 3 driver can be used with JDBC 1.2, JDBC 2.0, and JDBC 2.1.

The JDBC Type 4 driver is also a pure Java implementation that is just called a JDBC Type 4 driver for now, because there is no package for it in DB2 Version 7. DB2 Version 7 does not support a JDBC Type 4 driver. In an upcoming release of DB2 you can expect to see this support. An application using a JDBC Type 4 driver does not need to interface with a DB2 client for connectivity because this driver comes with Distributed Relational Database Architecture Application Requester (DRDA AR) functionality built into the driver.
Because a JDBC Type 4 driver supports DRDA and can make direct connections to a database, this may give rise to the notion that there will no longer be a need for DB2 Connect (the data management middleware that provides DRDA AR functionality to workstations). This simply is not true for many reasons (beyond the scope of this book). DB2 Connect is a comprehensive middleware solution that adds significant business value to an organization. DB2 Connect adds so much more to a solution, like J2EE-compliance (a JDBC Type 4 Driver is not J2EE-compliant), a support for not only JDBC, but ODBC, DB2 CLI, ADO, or embedded SQL, a connection concentrator, and access to non-mainframe resources such as CICS, IMS, VSAM, MQSeries, and much, much more!

John V. McLean

You can learn more about the XML Extender at www.ibm.com/software/data/db2/extenders/xmlext or see a demonstration at www.ibm.com/netdata.boulder.ibm.com/wwwdemo/xmldemo/xmldemo.html.

45.0 DB2 UDB NT Version 7 and Up –

Installation

NT NT NT NT NT NT NT NT

You start the installation process with a CD. Here is a check list.

1. Is the CD for the installation of DB2 UDB (Universal Dada Base) on a NT Server?

2. Check the Version number is the one you want. (i.e. V5.2, V6.1, V7.1, V7.2, V8.1 or higher).

Check the type of edition of DB2 UDB you want to install (Personal, Work Group. Enterprise etc), this should be on the label of the CD.

45.0.1 New Technology (NT)

ACCESS	Assume that you have created a ODBC data source called XYZ that allows you to access a database called XYZ, Start up Access and create a database. Access defaults to the databases you create DB1 DB2 etc. Select the Data Source. Supply Server User Id and password. Access will look at those tables with a primary key. Once tables have been selected by access you can start manipulating the data in them.	
Add Space	ALTER TABLESPACE Userspace1 PREFETCHSIZE 64; ALTER TABLESPACE longspace ADD (FILE 'f:\ longspace\space2.dat' 65000);	After a tablespace has been created the only alterations that can be made to it are as follows:- The PREFETCH, OVERHEAD, TRANSFERRATE and BUFFERPOOL properties of the tablespace can be changed as Shown. New containers can be added to a DMS tablespace as in the example shown. When a new container is added to a tablespace, a background process automatically moves some data from existing containers into the new container so that the data in the tablespace will be balanced across all its containers.
Alert Center	You can access the Alert Center by using the buttons on the Control Center tool bar. From the Alert Center you can see any alarms and warnings which have been generated by the Snapshot Monitor. If the Snapshot Monitor is set to issue a warning when a parameter reaches a certain level, that warning will appear in the Alert Center. It gives you early notification of potential problems. Colored icons represent different levels of warning, using the traffic light red – amber – and green color scheme.	
Aliases	An alias lets you use a different name for a database on the local machine. If you have a database ABC on the server, you can create an alias for it (example:- XYZ) on the workstation. In many cases this simply adds complexity.	

Backup	BACKUP DATABASE database name TABLESPACE tablespace name ONLINE TO e:\backups;	Backup all your data on a regular basis. Define a backup strategy, and follow the plan. Backups must also cater for disaster and disaster recovery.
Backups	There are two type of backups:- Off Line and On-Line Off Line:- This is the most straightforward type of backup. Ensure that no one is accessing the database and then make a copy. On Line:- Make a backup while Users are using the data base backup wizards. Sounds attractive, but more care needs to be taken. DB2 has Smart Guides for both backup and restore. Backup using SmartGuide Make sure no one is using the database. Go to the Control Center right click on the DB2 Instance and Select "Force Applications". This forces all applications for that instance off the system. Right click on the database and select "Backup Database Using SmartGuide" The SmartGuide opens at the Database page. This section lets you choose a database, but since you right clicked on the database, that is selected by default. Next select Availability:- Frequency. Next – Protection:- Select Complete Recovery with Archive Logging. Next – Rate of Change – What percentage of your data is new or changes each day. Next – Choose Archive Logging and Offline (this is safer). Next - Schedule – Every day at what time?? Midnight?? Next – add base image – where do you want to store it. Add the target location. The Wizard supports disk and tape backup on NT. Do not backup to the same disk holding the data. Press done and the Wizard will do the rest. The SmartGuide Generates a Script which you will find by opening the Script Center. This script has also been scheduled for execution and you can check this in the Journal. You can force a backup from either the Journal or the Script Center. So open one of them up, right click the backup job and select RUN. After a brief pause, your backup job should be visible in the Job History panel.	

Buffer Pools	A bufferpool is an area of Main Memory into which database pages are read and held during processing. You can tune the performance of your system by controlling the number and size of the bufferpools and the assignment of table spaces to bufferpools. The create action allows you to create a new bufferpool and to specify it's name and size. Alter bufferpool – allows you to change the size of the bufferpool. Drop bufferpool – this action drops the bufferpool.
CAE (Client Application Enabler)	No matter which version of UDB you install, the Client Application Enabler will be installed as well. The CAE is the minimum software a workstation needs to enable applications on the Client to talk to DB2. The CAE is a series of DLLs.
CCA (Client Configuration Assistant)	You can also install the Client Configuration Assistant (CCA) on a Workstation. The CCA can be used to set up and configure the links (typically ODBC) which enable front end communication with DB2. The CCA is automatically installed on the server for you when DB2 is installed. A user's workstation, one that is used solely for data access to a Database only needs CAE installed. However, your workstation needs more software installed upon it to test DB2 and play with the data it holds. You will almost certainly want a copy of the Control Center in addition you may want some documentation installed locally. IBM supplies a setup program that installs all of these pieces. The setup program is located on the Client CD-Rom. Pop the CD into your drive and install. Note:- You will need to be logged on to your work station with a Username that belongs to the Administrators Group for that Work station. You should also use a user name of eight characters or less and restricted in the same way as the User Id used to install UDB DB2 on the server.
Check Pending	If a loaded table has any foreign key or check constraints the Load utility leaves it in a Check Pending state. This means that constraints have not been checked for some of the data in the table. If the loaded table is a parent table in any referential integrity relationships, and if the load was done in replace mode, the child tables in these relationships are left in Check Pending state also. A table in check pending state cannot be accessed by SQL statements until some action has been taken to remove it from this special state. The usual way to remove a table from Check Pending state is by using the SET CONSTRAINTS statement. e.g. SET CONSTRAINTS FOR shop.tools IMMEDIATE CHECKED FOR EXCEPTION IN shop.tools USE shop.badtools; Immediate Checked = The bad rows are put in table shop.badtoools.

CLI (Call Level Interface)	Call Level Interface (CLI) has several important advantages that might cause you to prefer it for some applications over Embedded or Dynamic SQL or even static SQL. These advantages can be summarized as follows:- 1. CLI applications do not require a precompiler. CLI relies on a set of function calls. No need to bind either. 2. CLI enables you to write portable database applications. 3. CLI can connect to the same database multiple times and can independently commit transactions in each database Connection. 4. CLI provides a set of function Calls that can be used to access the System Catalog Tables. 5. CLI allows the application program to retrieve multiple rows from a result set with a single call. 6. CLI Supports scrollable cursors. 7. Supports a large number of datatype conversios. CLI depends on handles which is a C type long variable which represents some information. A configuration file named db2cli.ini can be used to specify various options that Control the behavior of CLI.
Connections	Before any access to data is possible, your application program or query interface must be connected to a database. In general your application may be running on a client machine, and the database may be located on a different server machine. During the course of an application session, it may be necessary to connect to more than one database. A database connection can be established explicitly by a SQL Connect statement such as:- CONNECT TO dbase1 USER auserid USING password You can also cause UDB to connect to a database automatically; this is done by setting the environment variable DB2DBDFT to the name of the default database you wish to use. If this environment variable is defined, DB2 UDB will establish an implicit connection to the default database you wish to use. There are two types of database connections:- Type 1 and Type 2. Type 1 – Each transaction in confined to a single database. Type 2 – A transaction may be connected to multiple databases possibly on different servers. The choice between Type 1 and Type 2 can be made by the SET CLIENT Command.
Containers	There are three types of containers:- 1. Files 2. Subdirectories 3. Areas of raw unformatted disk. DB2 can use two types of storage:- 1. SMS System Managed Storage (storage managed by the operating system). Expands and contracts as the data changes size. Easy to manage, but slow. 2. DMS Database Managed Space. Fixed size. You have to estimate the size of the tables when the database is created. FAST. File and Area of unformatted disk.

Control Center	Graphics available in the Control Center:-	
	Systems, Instances, Databases, Tables, Table Spaces, Views, Aliases, Schemas, Indexes, Triggers, Replication Sources, Buffer Pools, User Defined Types, User Defined Functions, Packages, Connections, DB Users, DB Groups, Nodes, Database Partitions, Table Space Partitions, Nodegroups.	
	You can find the control center by doing the following:- Start, Programs, DB2 for Windows NT, Administration Tools.	
	There are two windows in the Control Center, Left side is the Object tree, right is the contents pane, it displays the contents of the object selected in the left side.	
Create Instance	**db2icrt tstdb2**	Creates a new instance and gives it a name. In the example shown, the instance named tstdb2 is created. To create an instance in UNIX you must hold root authority. In NT you must be a member of the Administration Group. When you create an instance your group becomes the holder of SYSADM authority for the instance.
Databases	A database is a collection of related data and other components, such as tables and indexes. The data is held in one or more tables. Each instance of DB2 can manage one or more databases. Right Click and a menu appears which lists the operations appropriate for that type of object.	

DB Users	Any individual that accesses a database can be described as a User. There can be anything between 1 and thousands of Users on a System. A collection of Users can be defined as a Group. The Users in a group can be defined as it's membership. All DB2 Users have automatic membership of the predefined Group called PUBLIC. Groups often map neatly to the jobs that people do, for instance, into such divisions as programming, management, and sales. One of the advantages of Groups is that they make the allocation of privileges very easy. DB2 makes use of NT's Groups. Authorities There are four Authorities:- SYSADM – System Administration SYSCTRL – System Control SYSMAINT – System Maintenance DBADM – Database Administrator SYSADM – The highest level. As an installer of DB2 you automatically get this authority. You are all seeing and all powerful. You can access data in any database and have complete GRANT and REVOKE authority. You can do anything (you're God!). SYSCTRL – Only SYSADM can grant SYSCTRL authority. SYSCTRL has full System Authority. SYSCTRL is just like SYSADM but does not have the ability to read and modify data. SYSCTRL can create and drop instances and databases and even force Users off the System. SYSMAINT – Can only be granted by SYSADM. It is a lower level of authority than SYSCTRL. SYSMAINT can back up, roll forward recovery, start and stop a database, generally maintain an Instance or any of it's data bases. DBADM – is the fourth level of authority available. It can only be granted by SYSADM. DBADM applies to a specific data base and all the objects within it. Inspect log files, monitor events, reorganize tables, collect statistics. A DBADM can grant any privilege on any object within it's allowable database.

Drop an Instance	Db2idrop tstdb2	Drops (removes) an instance and all the databases that belong to it. This command requires the same level of authority as db2icrt. Db2idrop tstdb2 drops the instance named tstdb2.

Indexes	An index defines a primary unique key for your data which makes it's access rapid, otherwise DB2 would have to sequentially read through the rows until the relevant row was found. Type 2 indexes reduce locking.
Instance	Any machine that can run DB2 can be used to manage databases. However, it is possible to run multiple copies or instances of DB2 on the same server at the same time. An instance can be looked on as a Logical Server. The advantage of running multiple instances on the same machine is that different instances are completely independent and isolated from each other. You can run production and test databases off the same instance, or more than one DB2 test database off the same instance. Multiple instances of DB2 are used to provide a greater level of isolation between databases.

Journal	Once a script has been scheduled it is called a job. All jobs appear in the Journal. In the Journal you can see:- 1. All Jobs that are pending (waiting to execute) 2. All Jobs currently executing. 3. All Jobs that have completed. The Journal keeps track of time-based events. It's not just for jobs, For example, if you click the Recovery Tab and use the Select Button to locate a particular database, you can see the options for recovering the database from a backup at a particular time.	
List Instances	Db2ilist	Lists all the instances available on your system.
LOGS	Whenever a transaction starts against a database, DB2 records the fact in a log file. The log stores information like: 1. When a transaction starts. 2. What operations are completed. 3. When the transaction committed. <u>Circular Logging</u> The default type of logging is Circular Logging. This usually consists of three log files which are written to in a circular fashion. A fourth log file is known as a secondary log file. Circular logging enables DB2 to roll back any or all incomplete transactions. <u>Archive Logging</u> An Archived Log stores details of all transactions since the last backup and can be used to restore a database to any point between that backup and the failure. The log files are kept on disk in the database log path directory. By default this is set to the same disk as the database. You may want to move it to another disk. Be real careful and back up inactive logs as well. To configure backup parameters right click on database and select CONFIGURE.	

Monitoring DB2	DB2 provides several monitoring systems that will help you get the best from your system. These run in the background.
	<u>Event Monitoring</u> When an Event Monitor is running, data is collected in a continuous stream from events as they happen and is directed to a file. Once the monitoring is complete, the contents of the file are ready for inspection with the Event Analyzer. Event monitors work within a single database. They cover:- 1. database connections and disconnections 2. transactions 3. SQL statements 4. Deadlock activity To setup:- 1. Right click on a database object in the control center. 2. Choose Monitor Events 3. When blank Event Monitor window opens click on Event Monitor and choose Create. 4. Define the events you wish to monitor in the Event Monitor Window. Direct your output View the Monitor files. . You can define up to 32 event monitors at any one time <u>Snapshot Monitoring</u> Snapshot Monitoring is used to display information about a database In real-time. e.g. How many users are connected to the database at the same time. Right click on a table icon. Choose Snapshot Monitoring. Start Monitoring – Start actively monitoring this object. Stop Monitoring – Stop actively monitoring this object Show monitoring details – Show me details. Opens dialog box "Monitor Profile" Every performance variable is shown. Select tabs on the top for information about Database Instance, Database, Table, Table Space, Database Connections. You can set thresholds. Example:- If the number of reads exceeds 10 per second on this table space make a beeping sound and place a message in the Alert Center. You can also watch data as the data is collected and see how the values change over time. Note:- You can turn on monitoring for a specific object but turn off monitoring at the Instance Level.

Nodes	Inter Partitioned Parallelism was first introduced in DB2 Parallel Edition for AIX, in which each of the separate partitions of the database was called a node. In UDB, the term *partition* is introduced but some of the terminology from the original product is still in use. For example a group of partitions is called a *nodegroup* and is operated on by such commands as ALTER NODEGROUP. To avoid confusion remember that the terms node and partition are interchangeable.
	
	In UDB EEE you must create a node configuration file on each machine in the sqllib directory named db2nodes.cfg which lists the Host name of each participating machine and assigns a node number between 0 and 999 to each machine.
	e.g.
	    CREATE NODEGROUP group1 ON NODE (0);
	    CREATE NODEGROUP group2 ON NOES (1, 3, 5);
	    CREATE NODEGROUP GROUP3 ON ALL NODES;
	
	You can obtain a list of all nodes/partitions by:-
	
	LIST NODES
	LIST NODEGROUPS SHOW DETAIL;
	
	When a database is created the following default nodes are created:-
	
	IBMDEFAULTGROUP – serves as nodegroup for the table space
	        USERSPACE1
	IBMCATGROUP – serves as nodegroup for table space
	        SYSCATSPACE
	IBMTEMPGROUP – serves as nodegroup for table space
	        TEMPSPACE1

Packages	A list of packages that are bound in a given database can be found under Application Objects. Each packager is displayed along with it's authorization id., the time when it was bound, whether it is currently valid, and it's isolation level. "Show Explainable Statements" – Lists all the SQL Statements in the package for which an access plan is available for display. "Show Explainable Statements Hisotry" – Lists each time each statement was bound.

PARTITIONS	The concept of a partitioning key applies to your table, only if you create the table in a tablespace that is partitioned across multiple nodes in a parallel data base. If this is the case, the partitioning key controls the distribution of rows among nodes. If you create a table in a partitioned tablespace and do not specify a partitioning key, the default partitioning key is the first column of the primary key, if no primary key then the first column of the table is used as the partitioning key. For loading a partitioned database, UDB provides two utility programs called Splitter and Autoloader. The data must first be split into several subsets based on the value of a partitioning key, then each subset must be loaded into it's respective partition. Splitter is provided in the form of a program called DB2SPLIT, found in sqllib/bin and runs on all UDB platforms. It's job is to scan the data and prepare it for loading into a nodegroup in a partitioned database according to a partitioning key. The format can be in DEL, ASC, BIN or PACK. The Splitter is controlled by a configuration file which specifies names of the data files, nodes, partitioning key for each node and various other parameters. e.g. db2split –c db2split.cfg <u>Autoloader</u> The Autoloader is provided in the form of a program named db2autold, found in sqllib/misc that runs only on the UDB Enterprise Edition Extended. It provides an easy-to-use interface with the same functionality as the Splitter and with some additional Features as well. It can transfer its input data from a remote system via FTP, and after splitting the data it can actually load into a Designated node group. Like the Splitter it is also controlled by a Configuration file. Parms include:- Analyze, Split_only, Load_Only, Split_and_Load. e.g. db2autold –c autoloader.cfg
RECOVERY (Restore)	The Restore SmartGuide helps you restore a database. In the Control Center, right click the database to be restored and select Restore Database Using SmartGuide. You can choose from which backup to restore. Typically it's the most recent. Select the Roll Forward tab. Select Roll forward to the end of the logs. Click Done. The database is restored.

Reorg	REORG TABLE table name INDEX index name USE tablespace name	Reorg reorganizes the table space so that the pages are arranged in contiguous order.
Reorg check	REORGCHK ON TABLE table name	Reorg check lets you know if a table needs to be reorganized

REORGCHK	UDB provides a command called REORGCHK to help you decide When a table needs reorganizing. REORGCHK also updates statistical information. REORGCHK ON TABLE company.parts The table must always be qualified by a schema name. The example above generates a report on the physical organization Of the table called. If you invoke REORGCHK without any operands, it generates a report on the physical organization of all the tables that you are permitted to reorganize. E.g. REORGCHK For each table three formulas are computed:- F1 = Number of overflow records. F2 = Amount of free space on the pages where the table is stored. F3 = Number of empty pages that would be fetched in a scan of the Table. For each index of a table three formulas are also computed:- F4 = Measures the clustering property of an index. F5 = Measures the amount of free space on index pages. F6 = Whether the index has the appropriate number of levels. An * in any (F) column means that the particular item needs attention.

Replication	When you select replication sources, the control center displays a list of tables that are available as sources for data replication. You can select one or more of these tables and define a subscription. For replicating updates to one of more target tables, possible in a different database. The "define subscription" action causes a dialog box to appear in which you can give the subscription a name, specify the source and target tables, and specify how often you want the updates to be propagated. You also need to specify the Database in which the subscription is stored. The subscription then Becomes an object in it's own right and appears n the Control Center hierarchy under Replication Subscription for the database containing the source tables. For subscriptions you can select these actions:- Change, Clone, Activate, Deactivate, Remove. Subscriptions are processed by a Capture program that runs on the source data base and an Apply program that runs on the target database.

Restore	RESTORE DATABASE database name FROM e:\backups TAKEN AT 20021218150345 REPLACE EXISTING	RESTORE restores the content of a database using the content that was saved in a backup. REPLACE EXISTING means that the existing content of the database will be deleted and replaced by the content of the backup.
Runstats	RUNSTATS ON TABLE table name WITH DISTRIBUTION ABD DETAILED INDEXES ALL SHRLEVEL CHANGE	Always run RUNSTATS after REORG so that the CATALOG tables statistics are updated. Then do a REBIND any application programs that operate on the REORGed table to take advantage of the new organization.

Schemas	Every object in the data base belongs to a schema. Schemas are by default named after the person who creates the object. SYSADM can create schemas that have names other than your user id. All you have to do is right click on schemas in the object tree and type in the name. A schema name must begin with a letter and be eight characters or fewer in length. It must not begin with SYS and must not already exist. Schemas make an object name unique. Only if an object has your schema you can drop it. Otherwise SYSADM has to drop the object.
Script	This tool lets you manipulate scripts. It also allows you to schedule Scripts to run at particular times, run scripts immediately and Generally maintain them. In fact, when you create and save a script in the Command Center, it is automatically popped into the Script Center. Scripts can also come from other places. For example, when you use the backup Wizard. The Wizard automatically generates a script which is sent to the Script Center. So even if you never hand-craft a script you are still very likely to manipulate scripts and the Script Center is where you will do it. Note:- When you decide to schedule a script to be run at a certain time, the script automatically appears in the Journal Center.
Systems	The Systems that show up in the Control Center are simply DB2 Servers. If you are connected to a bigger network, other DB2 Servers may be visible to your server. If so they will appear in addition to the one you have just installed.
Table	Tables are allocated to tablespaces. Right click on Tables in the Object tree, make sure you are in the right data base. Select Create Table using smart guide. Smart guide has five tabs. Enter the table name, leave the table schema if you have previously set it up. Build the table columns, primary key, allocate the table space.
Tablespaces	Each table has to be assigned to a container and when the numbers grow, keeping track of which table is stored in which container can become a major pain. The concept of table spaces are simply groups of tables. You can allocate multiple tables to table spaces and multiple containers to table spaces. One absolute rule is that any given container is only ever allocated to a single tablespace. You are not allowed to allocate containers of both types to a single tablespace. Therefore tablespaces are either assigned with SMS or DMS containers never both. Tablespaces can be classified in another way into three types:- 1. Regular, 2. Long, 3. Temporary.

Triggers	A trigger can be thought of as a small program that is set to go off under certain conditions. A trigger is activated by three types of modification to a table. Namely the SQL statements:- INSERT DELETE UPDATE Triggers can be used for many things but are often used for data integrity checking. There are before and after triggers. e.g BEFORE UPDATE and AFTER UPDATE. e.g. A trigger can be set to notify if a new hire has a salary above a certain level. An event sets off a Trigger. A trigger is tied to a table. Right click on trigger to create a trigger and trigger dialog. A trigger id is the same as your schema.
Tuning	<u>EXPLAIN</u> Before you can use explain, you need to create Explain tables that are used to capture details of access plans. You can do this by connecting to the desired database and executing CREATE TABLE statements in the file sqllib/misc/EXPLAIN.DDL By default, the schema of the Explain Tables is the same as your userid. Thus, each user gets a separate set of Explain Tables. e.g. db2 connect to testdb db2 –tf EXPLAIN.DDL Creates the Explain tables in a database named TESTDB. EXPLAIN_INSTANCE Table:- Each row represents a package in which one or more statements have been explained. EXPLAIN_STATEMENT Table:- Each row represents an SQL statement that has been explained. EXPLAIN_OBJECT Table:- Each row represents the sources of Data used by a plan. Such as temporary or permanent table or index. EXPLAIN_OPERATOR Table:- Each row represents an operation such as 'Union" or "Merge Join" that was selected by the optimizer As part of a Plan. EXPLAIN_ARGUMENT Table:- The rows of this table contain Detailed information about the individual operators in a plan, such as columns used for sorting, duplicate values etc. REORG – See REORG
User Defined Functions	A list of User Defined Functions can be found in the Control Center Hierarchy under Application Objects. Each function is displayed with it's fully qualified name and specific name, parameter types and result type.
User Defined Types	A list of user defined types can be found in the Control Center hierarchy under Application Objects. You can create a new distinct type by invoking the Create Action. Each distinct type is displayed along with it's source type. You can drop distinct types if you wish.

Views	A view lets you provide users with a constrained image of a table. You can have many views of the same table. Sometimes used for security purposes. Lets users see only what they have to see. Right click, enter view name and SQL "As select Col1, col2 etc. from table Name. A view can be updated, where as a results set of a query cannot.

46.0 Miscellaneous

46.0.1 DB2 UDB Create SMS/DMS Tablespaces

SMS

In order to create an SMS tablespace, you need to specify the name of the tablespace and the path of all it's containers. Remember, each container in an SMS tablespace is a directory. If a relative path name is given, it is interpreted with respect to the local database directory of the database containing the tablespace. If the last level of the path name does not exist, a directory with a given name is created.

In the following example we create an SMS tablespace with four containers, implemented by directories on four different disks in an NT file system. We then create a table in the new tablespace, causing the data in the table to be distributed among the four disks. Before executing the CREATE TABLESPACE statement, our session must be connected to the database in which the tablespace is to be created.

Example:-

```
                              (Id.)              (p/w)
CONNECT TO database USER xxxxxxxx USING yyyyyyyy;

CREATE TABLESPACE sms1 MANAGED BY SYSTEM
 USING ('d:\sms1', 'e:\sms1', 'f:\sms1', 'g:\sms1');

CREATE TABLE accounts.receivable
                (custno       Char(10),
                 amount       Decimal(10,2),
                 dueDate      Date,
                 PRIMARY KEY(custno, dueDate)
                 IN sms1;
```

DMS

When creating a DMS tablespace, it is necessary to specify not only the names of the files or devices that implement the containers, but also their sizes. The size of each file s specified as a number of 4K-byte pages, and the specified amount of space is allocated when the tablespace is created. In the following example, we create a DMS tablespace consisting of two containers, each implemented by a file of 10,000 4K Pages.

<u>Example:-</u>

```
CREATE TABLESPACE dms2 MANAGED BY DATABASE
  USING (FILE 'd:\dms2\dms2.dat' 10000,
         FILE 'e:\dms2\dms2.dat' 10000;
```

UDB recognizes two specialized kinds of tablespaces called *temporary* and *long* tablespaces. A temporary tablespace provides space for the system to use for temporary results, such as a table that is materialized for sorting during the processing of a query. Every database must have at least one temporary tablespace at the time of it's creation by specifying CREATE TEMPORARY TABLESPACE.

A *long* tablespace is a tablespace that is dedicated to the storage of large objects (LOBs). When a table is created, the CREATE TABLE statement can specify whether the table will store it's large objects mixed with it's other data in a regular tablespace or use a separate long tablespace for this purpose. Separating the large objects from the rest of the data can improve the clustering properties of the table and reduce the number of I/O operations needed to scan the table, **this is strongly advised.** A long tablespace must be defined as a DMS (Database Managed Space) tablespace and is designated at creation time by specifying CREATE LONG TABLESPACE. In the following example we allocate a 65,000 page file to serve as a long tablespace and create a table that uses the long tablespace to store it's large objects:-

<u>Other Examples:-</u>

<u>Large Object - Long Tablespace</u>

```
CREATE LONG TABLESPACE longtbspace MANAGED BY DATABASE
      USING (FILE 'd:\longtbspace\space1.dat' 65000);

CREATE TABLE bridges
      (name                      Varchar(50),
       latitude                  double,
       longitude                 double,
       photo                     blob(1M))
       IN dms2        LONG IN longtbspace;
```

<u>Regular Tablespace</u>

```
CREATE TABLESPACE regular1 MANAGED BY DATABASE
      USING (FILE 'd:\regularone\space2.dat' 150000);

CREATE TABLE bridges
      (name                          Varchar(32),
       latitude                      double,
       longitude                     double,
       IN dms2 regular1;
```

A CREATE TABLESPACE statement has the following optional parameters:-

EXTENT SIZE specifies the unit of space allocation inside the containers of the tablespace in 4K pages.

PREFETCH SIZE specifies the number of pages to be fetched from the tablespace in advance of being referenced, in an attempt to anticipate page references and reduce waiting for I/O.

Prefetching of pages is done automatically when the system is scanning the whole table, or whenever it detects a regular pattern of pages being fetched from disk into main memory.

OVERHEAD is an estimate of the average latency time (in milliseconds) to begin a new I/O Operation in the tablespace and is provided as information for the SQL Optimizer.

TRANSFRRATE is an estimate of the time required in milliseconds to read one 4K byte page in the tablespace and is provided as information for the SQL optimizer.

BUFFERPOOL names the bufferpool to be used for fetching pages from disk for this tablespace.

IN NODEGROUP is an option that is meaningful only for parallel databases. It names the nodegroup on which the tablespace is stored. A nodegroup is a collection of nodes in a parallel system.

46.0.2 Create Function

Features:- External Scalar Function Statement - Arguments

CREATE FUNCTION func-name
Func-name - This is the name by which the function will be
invoked. It may include a Schema name. Like Trigo.tangent. The
schema name may not begin with the letters SYS.

Datatype - The type of the data i.e Varchar(25), Blob(32K) the
maximum length of the datatype must be explicitly stated. If
the datatype is a distinct type it is converted to it's source
type before being passed to the external function.

AS LOCATOR - The system will not allocate a buffer for the
return value, instead, it is the responsibility of the function
to create and return a locator that represents the return
value. E.g. Only used with Unfenced functions which make them
relatively dangerous to use. E.g. sqlduf length),
sqlduf_substr (designated sub string of locator value),
sqludf_create_locator (creates new empty locator), _append
appends some data from a buffer to the end of the object
represented by the locator), sqludf_free_locator (allows the
system to free up resources associated with a locator that
are no longer needed).

RETURNS - This clause specifies the result type of your functions.
The result type can be a built-in or a user defined datatype.

datatype CAST FROM - The datatype from which this datatype is
CAST.

SPECIFIC specific name - This clause gives a specific name to the
function to identifiy it among all the function instances with
the same function name.

EXTERNAL NAME implementation-name - This clause identifies your
function as an External function and tells the system how to
find the C function that serves as it's implementation. This C
function must be compiled, linked and placed in a directory on
the server machine from which it can be dynamically loaded by
the database system when needed.e.g. EXTERNAL NAME '/u/dbfns/
bin/mortgage!payment'
If no path name is specified, the system looks for your function
in the sqllib/function directory associated with your database.

Thus, if you put the mortgage binary file in sqllib/function you can shorten the phrase above to:-

EXTERNAL NAME 'mortgage!payment'

DETERMINISTIC or NOT DETERMINISTIC – NOT DETERMINISTIC means that your function might return different results from two calls with the same parameters e.g. from the random number generator. The word DETERMINISTIC is part of the SQL3 Standard.

NO EXTERNAL ACTION or EXTERNAL ACTION – This clause is also mandatory and it specifies whether your function performs some action that affects the world outside the database. For example you might write a function that sends mail to someone, writes into a file, or sets off an alarm.

FENCED or NOT FENCED – The FENCED option specifies that your function must always be run in an address space that is separate from the database. This option causes a performance penalty due to process switching when the function is called but protects the integrity of the database against accidental or malicious damage. An unfenced function runs in the same address space as the database and can damage the integrity of your data.

NOT NULL CALL / NULL CALL – This clause controls what happens when your function is invoked with a null value. If you specify NOT NULL CALL the system will never pass a null argument to your function, instead, if a null argument is detected the System will automatically consider the result of your function to be NULL.

LANGUAGE C or JAVA or OLE – This clause is mandatory and specifies the programming language in which your function is implemented. This determines the Linkage convention used by UDB for invoking the function.

NO SQL – This mandatory clause specifies that your external function contains no SQL statements. At present external functions are not allowed to access the Database.

PARAMETER STYLE DB2SQL or DB2GENERAL – This clause is mandatory:-

If LANGUAGE = C or OLE, PARAMETER STYLE = DB2SQL, If LANGUAGE = JAVA, PARAMETER STYLE = DB2GENERAL.

NO SCRATCHPAD or SCRATCHPAD - If the function is created
with the SCRATCHPAD Option, the function is provided with
a "scratchpad" area in memory that it can use to preserve
information from one function invocation to the next.

NO FINAL CALL or FINAL CALL - When a function is used in a
SQL statement the function may be called multiple times during
the processing of the statement. With the FINAL CALL option,
the function is called one extra time at the at the end of
processing the SQL statement.

ALLOW PARALLEL or DISALLOW PARALLEL - This clause tells the
system whether it is safe to execute the function in parallel
on multiple processors.

NO DBINFO or DBINFO - This optional clause causes UDB to pass
an extra parameter to the function containing a pointer to a
data structure containing information such as the name of the
current database, the name of the authid, the name of a table
and column,

The following clauses are required for an external function:-

RETURNS, EXTERNAL, DETERMINISTIC OR NOT DETERMINISTIC,

EXTERNAL OR NOT EXTERNAL, LANGUAGE, PARAMETER STYLE DB2SQL,

NO SQL

Example:-

```
CREATE FUNCTION addweeks(Date, Integer)
RETURNS Date
EXTERNAL NAME 'datefns!addweeks'
DETERMINISTIC
NO EXTERNAL ACTION
NULL CALL
LANGUAGE C
PARAMETER STYLE DB2SQL
NO SQL;
```

46.0.3 Unique UNIX Problem

The problem occurred when trying to LOAD/REPLACE a table from a previously created file via EXPORT.

The error message:- SQL2036N The path for the file or device "server name/instance/database/filename" is not valid.

Resolution:- Change the permissions on the exported file to "chmod 775", change all the directories (the one it was located in and the directories above it with "chmod 771").

Explanation:- Since all load processes (and all server processes in general) are owned by the instance owner, and all of these processes use the identification of the instance owner to access needed files, the instance owner must have read access to input data files. These input data files must be readable by the instance owner, regardless of who invokes the command.

46.0.4 Unix Functions / Actions / Comments

Function	Action	Comments
# times instance being used	ps –aef | grep –I xxxx | ws –I	Where I = lc and xxxx = last 4 bytes of instance
Alter the value of the NPAGES parameter in IBMDEFAULTBP to –1.	Db2 alter bufferpool ibmdefaultbp size –I	The NPAGES parameter of IBMDEFAULTBP is used as the bufferpool size unless it is –1, in which case the BUFFPAGE size is used.
Append to an already existing file	File 2 >> File 1	File 2 is added after File 1 in File 1.
Append to End of data line	$ a xx <esc> <shift> :wq (where xx equals two characters to be appended to end of line)	
Automatic scheduling	Crontab –e	Gets you into the Area where you can Set auto kickoff of Backups, runstats, etc.
Background Execution $ = Prompt	$ at now <enter> <enter command here> <enter> cntrl-D	You can substitute the 24 hour clock time for now e.g. 18:30 means your job will kick off at 6:00 PM
Cancel (Kill) a job	Ps <enter> Kill xxxxxxxx	1. To get Process Id# (PID) 2. Cancel Where xxxxxxxx = PID
Cancel a Command	<ctrl> -U or <ctrl> –X, or @	
Catalog DB to Node	DB2 catalog db Sample at node db2do01 (try in CLP)	Db2do01 = Instance name
Catalog to Node (Instance) – make sure your are in the system	DB2 Catalog db dwisq001 at node db2qo01 (in unix always say DB2)	Catalogs the database to the node
Change a string and copy file to new file.	Sed 's/chairman/chairbeing/' minutes > new-minutes	Change chairman to Chairbeing Make new file called New minutes
Change all	:1,$s/xxxx/yyyy/g	Change all xxxx to Yyyy globally
Change the working directory	cd /xxx	
Change working directory back one node	Cd	Lower Case
Check the BUFFPAGE size of the database	Db2 get db cfg for xxxxxxxx | grep BUFF	Where xxxxxxxx is the name of the database
Check the size of the NPAGES parameter in IBMDEFAULT	Db2 "select * from syscat.bufferpools'	This should give you the BPNAME, BUFFERPOOLID, NPAGES amount and PAGESIZE
Check what is running	Ps	Ps gives you the process ids. Use these PIDs to Kill
Combine two files	cat file1 file2 > newfile	
Command Options (single letter Commands)	Db2 list command options	All lower case
Concatenate data to Existing records. Concat with date at end	:1,$s/$/xxxxxxxxxxxxxx/g :1,$s/$/xxxxx01\/01\/0001/	The second lowercase 's' means concatenate From existing EOL
Concatenate date to existing records	:1,$s/$/mm\/dd\/ccyy/	Note '\' as date separators
Connect to instance	/export/home/xxxxxxxx/sqllib/db2profile	blank before /export
Connect to the database	Db2 connect to xxxxxxxx	Where xxxxxxxx is the name of the database.

Conversion from Binary to ASCII	Convto a 630 infilename outfilename.asc	On one line. 630 = Record length a = ASCII
Copy 20 records from One file to another	Head –20 file1 > file2	Lower case
Copy a File	cp oldfile newfile	Lower Case
Copy all files in this directory and in all its sub directories	cp –r * /newdir	
Copy file to another DIR zz	cp oldfile dirname	Lower Case
Copy Lines (yank put)	Yyp	Copies the line and Pastes it below the Line copied
Copy many files to Another directory	cp budget* dirname	Budget is generic * is wild card
CPU usage by Process	/usr/local/bin/top | more	/usr/local/bin = dir vi top
Create File	vi myfile.xxx	Makes new file or Retrieves existing Period
Current line number		
Cursor on Line #	<ctrl> G	
DB2 List tablespaces	Db2 list tablespace show detail	EEE by partitions eee = all
Define a ksh file:- on first line:→	# /usr/bin/ksh	Then follow with SQL etc
Delete 1 character	X (move cursor to the beginning of the text you want to delete.	Lower Case x
Delete 5 characters	xxxxx	Lower Case
Delete a Block of Lines	10dd	Will delete 10 lines
Delete entire line	Dd	Lower Case
Delete from Cursor to End of Line	D	Upper Case
Display a file after Copying it	Cat file1 >> file2	Lower Case
Display a file on the Screen	Cat filename	Lower case
Display a large file	Cat longfile.name | more	Lower case
Display current Dir	pwd	Lower case
Display Instance	Db2 get instance	Display the current instance
Display Job SYSOUT Messages	1. Cd /u/yourid/lst 2. ls –lt | more 3. vi xxxxxxx.yyyyyy.sysout	First get to local .lst directory
Display KSH Manual	man ksh	M in lower case
Erase a file	rm junkfile	No confirmation
Error or OK script	If [$? –ne 0] then Echo "n Error Message. \n" Else Echo " OK. Message \n"s	In English: If return code not equal 0 Then error message Else OK. Msg
Execute (Run) a batch job	nohup XXXXXXXX.sss &	Do not forget the & Output goes to nohup.out
Execute (Run) a Job Online (do only for very Small jobs.	XXXXXXXX.sss	Job Script only and hit enter
Execute process in background	At now	If you logout task will still execute
Execute SQL	db2 –tvf in-filename > out-filename	Results are in out-filename
Execute SQL in a file	Db2 –tvf in-file >> out-file	Results are in out-file
Execute, Read and Write Permissions	chmod 755 filename.xxx	Grant Execute, Read and Write permissions on file
Find Directories	Ls –l | more	A "d" in the left most byte s a directroy.
Find Name of File	Du –a /home | sort –n | tail -20	Find the 20 largest files beginning with /home

Find string when editing	/xxxxxxxxxx	Forward / then Immediately followed by string
Force Application	Db2 force application [ALL	(application handle)]	Use to break connections
Go down n lines stay in current column	Nj	Lower case n = #lines
Go to	NG	Goes to line n
Go to first non white space on preceding line	-	
Go to line at bottom of screen	L	Upper case
Go to line at top of screen	H	Upper case
Go to line in middle of screen	M	Upper case
Go up n lines stay in current column	Nk	Lower case n = #lines
Grep - Find a string in a file	Grep xxxxxx file1	All lower case xxxxx is the string
Group Check	View /etc/group	Shows id to groups
Insert text in front of the Current cursor position	I	Lower Case i
Instance connected to Process	Ps –aef	grep 25731	Where 25731 = process #
Jobs running under your Id. (To Find)	ps –ef	grep tmurphy	Where tmurphy is Your Id.
Jobs that are running under your id (look at).	Running xxxxxxxxx	All lower case where xxxxxxxxx Is your id.
KILL	Kill –9 xxxxxxxx	Use only if normal Kill does not work
List DCS	List DCS Directory	
List admin node	List [admin] node directory [show detail]	
List Applications	Db2 list applications for database Db_alias show detail	Displays entry for each application connected to a database
List –c Options	Db2 ? options	D = lower case
List command options – get to db2 Prompt first	?	List three ?s
List database directory	Db2 list database directory	grep –I alias	-I in lower case
List db2 commands	? db2-command	
List db2 help	? db2 help	
List DB2 Options	Db2 list options	D = lower case
List db2 options	? db2-options	
List File Names	Ls	
List file Names with sizes and dates	Ls –l	-L lower case
List Files and pause when screen is full	Ls –lt	more	List and sort by date/time most recent
List hidden files too	Ls –al	
List home directory	Ls –l /export/home	more	L = lower case
List Indoubt	List DRDA Indoubt transactions with prompting	
List Node Directory	Db2 list node directory	more	Node = Instance
List node directory	DB2 List Node Directory	more	Lists the databases in the nodes
List Nodegroups	List Nodegroups [show detail]	
List nodes	List nodes	
List packages	List packages [for {user	all	system	schema schema-name}]	
List selected names with Wild card	Ls *C7Z*	Where C7Z could be any characters
List tables	List tables [for {user	all	system	schema schema-name}] [show detail]	
List Tablespace Containers	List Tablespace Containers for tablespace-id [show detail]	
List tablespaces	List Tablespaces [show detail]	
List the List Commands	? List	
List/Get database Config Parameters	DB2 GET DB CFG	Can also be used when spelled out
Look at a long text file	more letter	
Look at a text file	cat letter	
Look at the manual page for the ls command	man ls	more	
Look1 db2look copies the DDLs from one database to another including privileges and views	Db2look –d dtoaq001 –a –e –l –x –f dtoaq011.ddl	Where dtoaq001 is the source db and dtoaq011 is the target db ddls.

Look2 After running db2look, excute the output ddls	Db2 –tvf dtoaq011.ddl –z dtoaq011.out	Dtoaq011.out is the output message file form the execution of the ddls.
LS Wild Card search	ls *xxxxxxxxx* or ls xxxxx*	Generic search of pwd library for item
Make a new name for a File (link)	ln oldname newname	
Make new name (links for files in a directory	ln dirname/* newdir	
Move 1 character to the Left	<esc> j	Lower case J
Move 1 character to the Right	<esc> l	Lower Case L
Move a file to another Directory	mv oldfile dirname/newfile	
Move cursor to beginning	0 (zero)	Move to beginning of current line
Move cursor to end	$ (dollar)	Move cursor to end of current line
Move to beginning of file	1G	Numeric 1
Move to end of line	G	Upper Case
Move to preceding line	-	Beginning of preceding line
New Command Line	<ctrl> C	
Number of words used	Concat.ksk	wc –l	Where concat.ksh Is a script
Password Change	Passwd	
Position Cursor at Column	N|	N = col# lower case
Print a file on a named printer	Lpr –Pprinter	
Print File	Lp textfile	Lower case
Print file	Lpr textfile	
Print file on named printer	Lp –dprinter	Lower Case
Print Job Cancel	Cancel requested or lprm job#	
Print only selected lines	Sed –n '/greasy/p restaurant-list	Search file restaurrant-list and print only those line Containing the string greasy
Printer Queue Check	Lpstat –a all or lpq –a	
Rename a file	mv oldfilename newfilename	
Repeat 10 lines	10yyp	From cursor repeat 10
Replace Text	R	
Restore Current Line	U	Upper case
Search for pattern	Search the text from where the cursor is looking for pattern	
Sort	Sort oldfilename > newfilename	Ascending order is the default. Left to right – whole record unless otherwise specified. Don't forget the >
Space used	Df -k .	Lower case df space period
Space Used	Df –k	more	List space used for All file systems
Space Used	Df –k name	Name = name of file system (Cash1d)
Table Copy	Db2look	Do ? db2look frist to get parameters
Telnet	To exit Telnet = cntrl-d	
Terminate the session	Db2 terminate	Terminates session
To edit	:e	Can edit a file
To increase table space size:- Error Msg:- The following xxxxxxx Tablespaces on yyyyyyy have exceeded maximum utilization percentages. Where xxxxxxx is the Instance and yyyyyyy is the server.	➢ db2 "alter tablespace zzzzzzz resize (file '/db2db/xxxxxxx/ dddddddd' 320000)" CAN USE EXTEND INSTEAD OF RESIZE Resize – makes file = to new size Extend – adds nnnnnn to current size	Zzzzzzz = tablespace name Ddddddd= database name 320000 = new size of the table space.
To quit and not save	<esc> :q!	

To Save Changes	<esc> :w	
To Save Changes and Exit	<esc> :wq or :zz	
Uncatalog database	Db2 uncatalog db sample at node db2do01	Db2do01 = Instance Name
Uncatalog Database	Db2 uncatalog database xxxxxxxx	
Uncatalog Node	Uncatalog node db2do01f	Uncatalog instance
Undo preceding Command	U	Lower Case u
UNIX FTP	Go to the directory where you would like to have the file sent. Type:- ftp 10.9.6.57 Type in user id. And password at the name prompt. Type:- pwd Do a cd to the sub-directory where The source data resides. Type:- get space source dsn space target dsn. You should hit enter and get a "File Transferred ". Type: -quit	Check target directory.
Verify the database name in the instance. First make sure you are in the instance.	Db2 list db directory | grep XXXX	Where XXXX are the first four characters of the database name in the instance to which you are working in.
Vi	View file-name	To look at a file – q! to exit
Vi	Save a file	:wq

47.0   REORGS

47.0.1   Reorganize tables used on the index.

By reorganizing the table on a commonly used grouping column(s)
or access column(s), the database manager already has the data
in the order it most often needs, thereby reducing the effort
required to return the data requested.

The Syntax is:-

REORG TABLE table-name INDEX index-name USE tablespace-name

47.0.2   Relation (Table) Scans

As the size of the table grows, the length of the actual time
taken to scan the entire table grows exponentially once the
size of the table exceeds the amount of memory available to
contain the pages from a relation scan. The significant increase
in time is due to page faulting in memory.

47.0.3   Data Distribution across Partitions

**Use the SELECT below to find out how many rows per table are
in each node of a partitioned Database:-**

```
------------------------------- Command entered ---------------
SELECT NODENUMBER(METCNCT_CUST_ID), COUNT(*) FROM TOLCUST_
CNTRC_NM GROUP BY NODENUMBER(METCNCT_CUST_ID)
---------------------------------------------------------------

         1                 2
----------- -----------
          3        6470908
          5        6405390
          7        6417207
          9        6392584
         11        6402600


  5 record(s) selected.
```

Where:- METCNCT_CUST_ID is the partitioning key.

47.0.4   Updateable Partitioning Keys

DB2 V7.2 allows you to update the columns in a partitioning key (a partitioning key exists only in a table spread over multiple database partitions).

Before DB2 V7.2, if you wanted to change the partitioning key, you needed to do two steps:-

1. Delete the row.
2. Insert the row with the new key.

Each step impacted the log space requirements on both the database partition losing the data and the one gaining the data.

With DB2 V7.2, this can be done in one step with an update statement.

In an Online Transaction Processing (OLTP) environment, updateable partitioning keys provide performance improvements in data redistribution.

To take advantage of this DB2 V7.2 enhancement, your JDBC Driver has to be at Version 2.0

48.0 DB2Batch

DB2Batch is a utility that gives statistics on the SQL Query . e.g. Elapsed Time, CPU Time etc.

db2batch -d dbname -f file_name [-a userid/passwd] [-t delcol] [-r outfile,[out file2]]
    [-c on/off] [-i short/long/complete] [-o options] [-v on/off]
    [-s on/off] [-q on/off/del] [-l x] [-cli [cache size]] [-h]

where: -f Input file containing SQL statements
    -d Database name. Default dbname set in $DB2DBDFT
    -t Single character column separator, except for tab column delimiter
        specify -t TAB for tab column delimiterts and equipment attached
    -a Authentication Host user id/password d therein are confidential,
    -r Output file containing query results. <outfile2>, if specified, will
        contain the summary table. Default is screen, duplication or
    -c Automatically commit after each SQL statement. Default is onempt
    -i Elapsed time interval measurementand/or to decrypt its contents,
        (Default) short   - time to open, fetch, and close cursoraw of
               long   - time up to start of next querythe right to
                    (includes db2batch overhead)systems by
        complete - time prepare, execute, and fetch separately
    -o Control options: all computer use, computer memory storage, and
        ie: -o r <rows out> (see details below)a computer."

DB2Batch Examples:-

```
db2 CONNECT TO database USER ..... USING ........

then:-

Key the following COMMAND SYNTAX in native UNIX/AIX or in CLP mode.

db2batch -d e8412pol -f sthis.sql -a e8412dba/OLCIUDB -r sthisout.txt -i
complete -o p 5

Where:-

db2batch:- invokes the bench mark testing tool.

-d:- is the identifier marker for the database name to follow.

e8412pol:- is the database name for which db2batch is being run

-f:- is the identifier marker for the input file containing the SQL
statements being analyzed.
```

-a:- is the identifier marker for the User Id. and Password.

e8412dba/OLCIUDB:- is the User Id and Password - Must be separated by a /

-r:- is the identifier marker for the output file.

sthisout.txt:- is the name of the output results file.

-i:- is the identifier marker for the elapsed time you want the tool (db2batch) to run.

Complete:- means you want to run to completion (short and long are the other options.

-o:- means you are going to use some options (these options follow the
     -o).

p:- is the identifier marker for the level of performance information returned.

5:- specifies that you want the highest level of performance
    information returned.

Example :- For testing the SQL related to a SBACCT_ALLOC table, make an input file called sbacct.sql and put the related SQL inside this file. Make a file called sbacctout.txt for the output. Leave all the other parameters the same.

49.0   Temporary Space

The TEMPSPACE1 table space is defined in the IBMTEMPGROUP Node Group.

If you do not specify any tablespace parameters with the Create Database command, the database manager will create these tablespaces using SMS Directory Containers. These Directory Containers will be created in the subdirectory created for the database, the extent size for these tablespaces will be set to the default.

The naming schema for Unix is:-

Specified_path/$DBINSTANCE/NODEnnnn/SQL00001

To create a temporary tablespace:- CREATE TEMPORARY TABLESPACE TEMPSPACE2 MANAGED BY SYSTEM USING ('d')

Then drop the old temporary tablespace supplied by the system:- DROP TABLESPACE TEMPSPACE1

Also you may try using DMS and specifying your own space parameters.

CREATE TEMPORARY TABLESPACE TMP#TEST
MANAGED BY SYSTEM USING ('E:\DB2\NODE0000\E8412TOL\TMP#REST');

DROP TABLESPACE TMP#TEST;

50.0 Partitioning Calculation

SELECT PARTITION(column name), COUNT(*) FROM table_name
GROUP BY PARTITION(Column_name)
ORDER BY PARTITION(Column name) DESC
FETCH FIRST 100 ROWS ONLY;

This SQL will give you the amount of rows in each partitioning map bucket from 0 to 4095 (4096).

It does not give you the number of rows of a specific table in a partitioned node.

51.0   REORGS

CLUSTERRATIO or normalized CLUSTERFACTOR (F4) will indicate
REORG is necessary for indexes that are not in the same sequence
as the base table. When multiple indexes are defined on a table,
one or more indexes may be flagged as needing REORG. Specify the
most important index for REORG sequencing.

52.0 Relation Table Scans

As the size of the table grows, the length of the actual time taken to scan the entire table grows exponentially once the size of the table exceeds the amount of memory available to contain the pages from a relation scan. The significant increase in time is due to page faulting in memory.

53.0 DB2ADVIS – Index Advisor

The Index Advisor is a tool that reduces the need for you to design and define suitable indexes for your data.

db2advis is good for:-1. Finding the best indexes for a problem query.

2. Finding the best indexes or a set of queries (workload) subject to resource limits.

3. Testing an index on a workload without having to create the index.

There are concepts associated with the SQL advise facility. First, there is a *workload*. A workload is a set of SQL statements which the database manager has to process over a given period of time. The SQL statements can include SELECT, INSERT, UPDATE and DELETE statements. For example, over a one month period of time, the database manager may have to process 10,000 INSERTS, 20,000 UPDATES, 10,000 SELECTS and 5,000 DELETES. The information in the workload is concerned with the type and frequency of the SQL statements over a given period of time. The 'advising engine' uses the workload information in conjunction with the database information to recommend indexes. The objective of the advising engine is to minimize the total workload cost.

Secondly, there is a concept of *virtual index*. Virtual indexes are indexes which do not exist in the current database schema. These indexes could be either recommendations that the Advise Facility has made to you, or indexes that you are looking to the Advise Facility to evaluate for you. These indexes could also be those the Advise Facility considers as part of the process and are passed back and forth from you to the Advise Facility using the ADVISE_INDEX Table.

53.0.1 How to use db2advis

Recap:- db2advis is a tool that recommends indices based on an existing database and a set of queries.

SYNTAX:

```
db2advis -d <db name> -t <time> -l <disk_space> -s "sql stmt"
-i <infile> -w <workload name> -o <output script> -p
```
 NOTE: only one of the following three options can be used:
[s,i,w]

where:
    -d database name.
    -p keep plans in explain tables.
    -t maximum duration for db2advis to run, in minutes.
       default is 1 minute, a value of 0 means unlimited
       duration.
    -l maximum disk space in megabytes. default is
       unlimited.
    -s recommend indexes for this SQL statement.
    -i get SQL from this input file.
    -o place the index creation script in a file.
    -w get SQL from rows in the ADVISE_WORKLOAD table,
       specified by matching WORKLOAD_NAME.

Example 1:-
 Create an input file called db2advis.in with the following
5 lines:-

```
--#SET FREQUENCY 100
SELECT COUNT(*) FROM EMPLOYEE;
SELECT * FROM EMPLOYEE WHERE LASTNAME='HAAS';
--#SET FREQUENCY 1
SELECT AVG(BONUS), AVG(SALARY) GROUP BY WORKDEPT ORDER BY
WORKDEPT;
```

Run the following command and let it finish:-

Connect to the sample database first
db2advis -d sample -i db2advis.in -t 5 -o db2advis.out
Example 2:-
        Connect to database e8412qol first
        db2advis -d e8412qol -t 0 -i dadvisin.sql -o
        dadvisout.txt

54.0 Configuration Parameters and the Path to the Logs

There are two levels of configuration parameters, one at the instance level (in this example initially connect to the instance exwintq007), and one at the database level, in this example connect to database e8412qol.

To display the configuration parameters at the instance level (a.k.a. the database configuration <u>manager)</u> enter:-

db2 get dbm cfg >> dbcfgmgr.out

Direct the output to an output file which you may browse later.

To display the configuration parameters at the database level (a.k.a. database configuration) enter:-

db2 get db cfg for e8412qol >> dbcfg.out

Direct the output to an output file which you may browse later. Note the database name is e8412qol.

If you want to see the path to the logs and the log files, look for "Path to log files" as the parameter prompt,
then change directories to the path to the logs. This information is contained in the database configuration
parameters. The first log file is at S0012825.LOG. As you can see permission was denied in this instance.

<u>From the database configuration parameters for e8412qol:-</u>

```
Log file size (4KB)                (LOGFILSIZ) = 2500
Number of primary log files        (LOGPRIMARY) = 10
Number of secondary log files      (LOGSECOND) = 20
Changed path to log files          (NEWLOGPATH) =
Path to log files                               = /prd/
d001n001log/udbqol01/e8412qol/NODE0001/
First active log file                           = S0012825.
LOG
```

Note:- In a partitioned database the NODE0001 is given as the default, other Node numbers as appropriate may be used for each partition.

55.0   Performance Tuning Check List

Nbr.	Item	Check
1	Bufferpool	Is the bufferpool the maximum it can be for the tables in the transaction under UDB V7.2?
2	Check Catalog Tables	Have the System Catalog tables been checked for each table in the transaction / database?
3	Clustered Index	Are the indexes belonging to the tables in the transaction Clustered?
4	Comparisons	When comparing the key columns of two tables, are the keys in the same order?
5	Configuration Parameters	Have the 'database' and 'database manager' (instance level) parameters been checked?
6	Constraints	What are the constraints on the tables involved in the transaction, are they required?
7	Co-related Names	When the table name is corelated, make sure all columns referenced start with the corelated name
8	Data Distribution	Are the data in the tables evenly distributed across all partitioned nodes. Use SQL to find out.
9	DB2batch	Has the bench mark tool db2batch been run and results analyzed on the transaction?
10	Db2look	Has db2look been used to capture statistical information about the tables?
11	DDLs	Have the DDLs been thoroughly checked?
12	Defaults	What are the various defaults for tablespace, bufferpool etc. Are defaults prevalent in the system?
13	Erwin	Has the DDL in use been forward engineered from the Erwin model??
14	Explain	Has an Explain been run on the SQL? Identify relation (table) scans and eliminate if possible.
15	Flat Files	Are flat files sorted in the desired order before loading to DB2 tables??
16	Optimization Class	Is the current setting of the optimization class adequate?
17	Partitioning Key	Is the partitioning key the first column in the relevant tables?
18	Referential Integrity	Is there any unnecessary referential integrity involved with the transaction.
19	Reorg	Has a recent reorg been run on the tables relating to a transaction?
20	Reorg Check	Run reorgchk to determine which tables need reorging – thus preventing unnecesary reorgs.
21	Reorg Index	Do the reorgs for tables specify the table name, index name and the tablespace name?
22	Row Blocking	Is row blocking used in Select statements e.g. OPTIMIZE FOR 20000 ROWS
23	Runstats	Have runstats been run immediately after the most recent reorg?
24	SQL	Is the SQL too complex? This may show in the Explain results where there are many table scans.
25	Table vs Index	Do the sequence of key fileds in the table match their definition in the index?
26	Tablespace	Have large tables been allocated DMS tablespaces? SMS is too slow to handle > 1 million rows
27	Timestamps	Are there any timestamps defined in the index columns of a talbe?? Get rid of them.
28	UDB V7.2	Is maximum use being made of the new features in V7.2? Example:- Temporary Tables? Partitioning Key Update?

56.0   Reorganize tables using the index.

By reorganizing the table on a commonly used grouping column(s)
or access column(s), the database manager already has the data
in the order it most often needs, thereby reducing the effort
required to return the data requested.

The Syntax is:-

REORG TABLE table-name INDEX index-name USE tablespace-name

Example:-

REORG TABLE E8412DBA.TOLST_HIST INDEX XOLCSTHS   USE TEMP01

Note:- If the Index is clustered, you do not need the index name in the REORG Statement

57.0 Bufferpools

The single most critical system related factor influencing DB2 performance is the setup of sufficient bufferpools. A bufferpool acts as a cache between DB2 and the physical DASD devices on which the data resides. After data has been read the Buffer Manager places the page into a bufferpool page stored in memory. Bufferpools therefore reduce the impact of I/O on the system by enabling DB2 to read and write data to memory locations synchronously, while performing time intensive physical I/O asynchronously.

By managing the bufferpools 'correctly', DB2 can keep the most recently used pages of data in memory so that they can be reused without incurring additional I/O. A page of data can remain in a bufferpool for quite a long time, as long as it is being accessed frequently.

How does the bufferpool work? DB2 performs all I/O related operations under the control of the Buffer Pool Manager (a sub set of the Database Manager). As pages are read from DASD, they are placed in 4K pages in the bufferpool using a hashing algorithm. The algorithm is based on an identifier for the data set and the number of the page in the data set. When data is subsequently requested, DB2 can check the bufferpool quickly using hashing techniques. This provides for efficient data retrieval. Additionally, DB2 data modification operations write to the bufferpool, which is far more efficient than writing directly to DASD.

How does DB2 keep track of what is updated in the bufferpool? This is accomplished by attaching a 'state' to each bufferpool page:-

1. available or 2. not available.

An available bufferpool page meets the following criteria:-

1. The page does not contain data updated by an SQL statement, which means that the page must be externalized to DASD before another page can take it's place.

2. The page does not contain data currently being used by a DB2 application.

An unavailable page is one that does not meet both of the criteria because it has either been updated and not yet written to DASD, or it is currently in use. When a page is available, it is said to be available for *stealing*. *Stealing* is the process whereby DB2 replaces the current data in a buffer page with a different page of data. The least recently used available buffer page is stolen first.

Although every shop's usage of bufferpools differs, some basic ideas can be used to separate different types of processing into disparate bufferpools. I list below one possible scenario of bufferpool usage:-

BP1   - Catalog and Directories
BP2   - Dedicate to sorting
BP3   - Sequential Index
BP4   - Sequential Tablespace
BP5   - Code and Lookup Tables
BP6   - Indexes for Code and Lookup Tables
BP7   - Dedicated bufferpool for an entire application
BP8   - Dedicated bufferpool for a single critical index
BP9   - Randomly accessed tablespace bufferpool
BP10   - Randomly accessed index bufferpool
BP11   - Reserved bufferpool for tuning and special testing.
BP12 up Additional bufferpools per tablespace, index, partition, application any combination thereof.

An often used technique for the use of bufferpools is to use separate bufferpools for indexes and tables. The idea behind this strategy is to enable DB2 to maintain more frequently accessed data by type of object. For instance, if the indexes are isolated in their own bufferpool, large sequential prefetch requests do not cause the indexes to be flushed, because the sequential prefetch is occurring in a different bufferpool. Thus, the index pages usually remain in memory longer, which increases access for indexed access.

You can further tune your bufferpool usage by isolating random access from sequential access. Doing this optimizes bufferpool usage so that each type of access that predominates is given it's own bufferpool. This increases performance.

Large (over 1 million rows) or very large tables should have their own tablespace and bufferpool, this will further improve performance.

Look up tables and code tables should have their own bufferpool, this further increasing performance.

One 4K bufferpool should always bee reserved for tuning and testing.

57.0.1 One large Bufferpool?

The general recommendation from consultants and IBM engineers in the past was to use only one large bufferpool or everything. This strategy turns over to DB2 the entire control of bufferpool management. The theory was that since DB2 uses efficient buffer handling techniques, good performance could be achieved using a single large bufferpool.

Today only a very few shops (most of them small) can get by with one large bufferpool. As the amount of stored data in DB2 databases increases, specialized types of tuning are necessary to optimize access. This usually results in the implementation of multiple bufferpools.

57.0.2 Multiple Bufferpools

Before implementing multiple bufferpools, be sure that your environment has the memory to back up the bufferpools. The specification of large bufferpools without sufficient memory to back them up can cause paging. Paging to DASD is very nasty and should be avoided at all costs. You should use the DB2 Catalog queries to keep track of multiple bufferpool usage.

57.0.3 Bufferpool Size

DB2 just loves large bufferpools! Each shop must determine the size of it's bufferpools based on the following factors:-

1. Size of the DB2 applications that must be processed.

2. Desired response time for DB2 applications.

3. Amount of real and virtual storage available.

But remember, DB2 does not allocate bufferpool pages in memory
until it needs them. A DB2 subsystem with very large bufferpools
might not use them most of the time.

As with the number of bufferpools, there are various schools of
thought on how to best determine the size of the bufferpool.
Try to allocate as large a bufferpool as possible within the
limitations defined by the amount of real and virtual memory
available.

The following calculation may be used as a rough starting point
for determining the size of your DB2 bufferpools:-

[number of concurrent users x 80] +
[(desired number of transactions per second) x (average
GETPAGEs per transaction)] +
[(Total # of leaf pages for all indexes) x 0.70]

The resulting number represents the number of 4K pages to
allocate for all your bufferpools. If you are using multiple
bufferpools then a percentage of this number must be apportioned
to each bufferpool you are using.

To determine the number of leaf pages for the indexes in your
DB2 subsystem, use the following query:-

```
SELECT SUM(NLEAF)
FROM SYSCAT.INDEXES;
```

Note:- RUNSTATS must be up to date for this query to be
meaningful.

58.0   Indexes

An index is a list of the locations of rows sorted by the contents of one or more specified columns. Indexes are typically used to improve query performance, however, they can also serve a logical data design purpose. For example, a unique index does not allow the entry of duplicate values in index columns, thereby guaranteeing that no rows of a table are identical. Indexes can be specified as being in ascending or descending order, ascending (ASC) is the default. The indexes contain a pointer known as the RID (record id) which points to the physical location of rows in a table.

More than one index can be defined on a particular table, which can have a beneficial effect on the performance of queries, however, the more indexes there are, the more the database manager must work in order to keep the indexes up-to-date during UPDATE,DELETE and INSERT operations.

A *Unique Index* guarantees uniqueness of the data values in a table's columns. The unique index can be used during query processing to perform faster retrieval of data. The uniqueness is enforced at the end of the SQL statement that updates rows or inserts new rows. The uniqueness is also checked during the execution of the CREATE INDEX statement. If the table already contains rows with duplicate key values, the index is not created.

A *unique key* is used to implement unique constraints. A unique constraint does not allow two different rows to have the same values in the key columns.

A *primary key* is used to implement entity integrity constraints. A primary key is a special type of unique key. There can only be one primary key per table. The primary key column(s) must be defined with the not null option.

A *foreign key* is used to implement referential integrity constraints. Referential constraints can only reference a primary key or unique constraint. The values of the foreign key can only have the values defined in the primary key. A foreign key is not an index.

The index supporting a primary key is known as the primary index of the indexes of the table. If a constraint name is not provided, DB2 will provide one for you, example: SYSIBM.SQL<timestamp>.

Indexes supporting primary or unique key constraints cannot be dropped explicitly. To remove primary or unique key constraints you need to use the ALTER TABLE statement. Primary Keys are dropped with the ALTER TABLE *table name* DROP PRIMARY KEY option. Unique key indexes are dropped using the ALTER TABLE DROP UNIQUE (*CONSTRAINT NAME*) option.

The Primary Key is a unique index and is used for data integrity and referential integrity verification. This cannot be done by just having a Unique Index, you must specify a Primary Key if you want a parent / child relationship. On the other hand, you cannot specify the CLUSTER, INCLUDE or PCTFREE parameters on the Primary Key definition. Therefore, under these circumstances, where you want a Primary Key and you want the unique index CLUSTERED, the use of the INCLUDE statement (more about this soon) and the PCTFREE option, you have to specify the Unique Index and the Primary Key separately in the same DDL even though you are referring to the same columns. DB2 will give you a Warning Message when you do this (words to the effect that "you already have a Primary Key"), but it is only a warning, and does not interfere with the implementation of the Unique Index with Options or the Primary Key. The meanings of the options specified on the Unique Index are as follows:-

CLUSTERED You can specify a clustering index that will be used both to cluster the rows during re-organization and to keep this characteristic during insert processing. Subsequent updates and inserts may make the index less well clustered (as measured by RUNSTATS) this resulting in periodic reorganizations. To reduce the frequency of reorganization you use the PCTFREE parameter (see below).

The degree to which the data is clustered with respect to the index can have a significant impact on performance and you should try to keep the unique index on the table close to 100% clustered. A clustered index attempts to maintain a particular order of data improving the CLUSTER RATIO and CLUSTERFACTOR statistics collected by the RUNSTATS utility.

Note:- A clustering index gives DB2 a way to look/scan over all the table rows while fetching the minimum number of physical pages because the indexes are in the order required.

PCTFREE This parameter tells the system to keep a specified percentage of free space on each page of the index when the index is created in order to allow for future inserts and updates. The default is 10%.

INCLUDE:- When you have a Unique Index, all the columns in the Index may not be used all the time, therefore for the least used column(s) you may INCLUDE them, meaning that they become part of the index only when invoked by the SQL. If they are not referred to in the SQL, then they are not considered part of the index. This improves performance by not letting the INCLUDEd column(s) participate in a comparison when they are not needed.

Example:- CREATE UNIQUE INDEX WITH THE INCLUDE, CLUSTER and PCTFREE parameters. Also use this as the PRIMARY KEY.

CREATE UNIQUE INDEX EDBA.XOLCSTHS ON EDBA.TOL _HIST (CNTRC_ ID, CNTRC_SFX_CD, SRC_SYS_CD, SEQ_NUM, METCNCT_CUST_ID, ST_ HIST_DIM_TS) INCLUDE (EFF_DT) CLUSTER PCTFREE 30 ;

ALTER TABLE EDBA.TOLST_HIST ADD CONSTRAINT XOLCSTHS PRIMARY KEY (CNTRC_ID, CNTRC_SFX_CD,SRC_SYS_CD,SEQ_NUM, METCNCT_CUST_ ID, ST_HIST_DIM_TS);

59.0 DB2 Performance

The following is a list of changes in alphabetical order that can affect DB2 performance.

Adding or removing indexes
Adding or removing memory
Adding or removing triggers
Adding rows to a table
Altering tables
Application level changes
Applying maintenance to DB2 software
Changing the clustering index
Changing the hardware environment
Compressing data
DB2 system level changes
Deleting rows from a table
Distributing data
Downsizing, upsizing, rightsizing
Ensuring proper sizing and quantity of log datasets
Enterprise wide changes
Implementing data sharing
Implementing Stored Procedures and / or User Defined Functions
Increasing system throughput
Increasing the application workload
Increasing the volume of inserts causing unclustered data or
   data set extents
Increasing the volume of updates to indexed columns
Incurring DB2 growth without resizing or reorganizing the
   catalog
Installing a new version or release of DB2
Integrating applications to the web
Key sequences not in the same order across the system
Modifying DB2 dispatching priorities
Moving data from site to site
Moving physical data to different volumes
Partitioning key not the first column in the table
Primary key greater than 40 bytes long
Rebinding
Reorganizing tables
Replicating and propagating data
Updating Runstats information
Using a varchar data type column as an index column

Note:- Thus, every change you make to any element of DB2 may have a profound effect on performance. It is a good idea to keep a log of every single change made against a performance chart(s).

59.0.1 The DB2 Optimizer

The DB2 Optimizer is referred to as a piece of "software", and it is generally accepted as such, but people get confused and even irritated when they are unable to directly "command" the Optimizer to do their bidding. This is because the Optimizer is not really software in the generally accepted sense of the word, but an intricate web of mathematical formulas and algorithms that emerge as one of the most complex pieces of "software" ever conceived. It shares this limelight with such other pieces of "software" as the Voyager Guidance System, the Hubbell Telescope, the Space Shuttle Systems and other very advanced projects. You may well ask "why did IBM do it this way?". The reason IBM chose mathematical formula to calculate access paths to data is three fold. First, the "foot print" of the Optimizer is the smallest on the market for the options it covers. Second is speed and third cost. The optimizer has the lowest cost per data byte accessed in the shortest space of time. No other RDBMS is as fast or as cheap as the DB2 Optimizer.

So if we can not order the optimizer to do our bidding, what can we do to get the results we want from a piece of SQL? Answer:- several things, known as "influencing the optimizer". Assuming you think you've done everything regarding space, indexes, table size, reorganizations and runstats, then try:-

Lowering the filter factor of a query by adding redundant predicates. Example: Change SELECT LASTNAME FROM TOLADR WHERE WORKDEPT = 'IT'
To
SELECT LAST NAME FROM TOLADR WHERE WORKDEPT = 'IT' AND WORKDEPT = 'SOFTWARE' AND WORKDEPT = 'DATA PROCESSING'

The redundant predicates are the last two 'where' statements and do not affect SQL statement functionality. However, DB2 calculates a lower filter factor, which increases the possibility that an index on the WORKDEPT column will be chosen. The lower filter factor also increases the possibility that the table will be chosen as the outer table, if the redundant predicates are used in a join (more about the filter factor later).

When redundant predicates are used to enhance performance, as detailed above, be sure to document the reason for the extra predicates. Failure to do so may cause somebody to assume they are an error and then remove them.

Another way to get a small performance boost from an SQL statement is to change the physical order of the predicates in the SQL code. DB2 evaluates predicates first by predicate type, then according to the order in which it encounters the predicates. The first four types are listed in the order DB2 processes them:- 1. Equality, 2. Ranges, 3. IN – where a column is tested against a list of values, 4. Place a predicate with fewer values in a table before a predicate with more values in the table, for example if there are more males in the entire company than workers in a specific department (which is very likely)
then in the SQL statement:- SELECT LASTNAME FROM TOLADR WHERE WORKDEPT = 'DATATEAM' AND SEX = 'M'.
Since there are less workers in the datateam than males in company, test the number of workers in the datateam first. This will shave a bit off the query's processing time.

59.0.2 Event Monitor for future use

59.0.3 The Filter Factor

The DB2 Optimizer calculates the filter factor for a query's predicates based on the number of rows that will be filtered out by the predicates.

The filter factor is a ratio that estimates I/O costs. The formulas and algorithms used by the Optimizer to calculate the filter factor are proprietary IBM information. These formulas assume uniform data distribution, so they are most accurate when determining the filter factor of static SQL queries, or queries on tables having no distribution statistics stored in the DB2 Catalog. The filter factor for dynamic SQL queries is calculated using the distribution statistics in SYSCAT. COLDIST, if available.

For example, consider the following query:-

```
SELECT EMPNO, LASTNAME, SEX
FROM TOLNAME
WHERE WORKDEPT = 'DATATEAM';
```

Assume the first column has an index called TOLNAME.XEMP2. If this query were being optimized by DB2, the filter factor for the WORKDEPT predicate would be calculated to estimate the number of I/Os needed to satisfy the request.

The link to the filter factor for this predicate is stored in column FIRSTKEYCARD in SYSCAT.INDEXES, say this number is 9. Then the filter factor for this query is 1/9 or .1111. In other words, DB2 assumes that approximately 11 percent of the rows from this table will satisfy this request.

You might be wondering how this information helps you; well, with a little bit of practical knowledge you can begin to determine how your SQL statements will perform before executing them. Remember:- the lower the % of the filter factor, the lower the cost and the more efficient your query will be. Thus, you can easily ascertain that as you further qualify a query with additional predicates you make it more efficient because the I/O requirements are reduced.

60.0   Table Spaces

60.0.1   Non Partitioned Tablespace SMS

```
CREATE REGULAR TABLESPACE SOL#ATRL IN NODEGROUP THREE_NG
PAGESIZE 4096 MANAGED BY SYSTEM USING ('/db2/d001n003di/
e8412iol/e8412dba/SOL#ATRL', '/db2/d002n003di/e8412iol/
e8412dba/SOL#ATRL', '/db2/d003n003di/e8412iol/e8412dba/
SOL#ATRL') ON NODES (3)
EXTENTSIZE 32
PREFETCHSIZE 96
BUFFERPOOL THREE_NG
OVERHEAD 24.100000
TRANSFERRATE 0.900000;
```

60.0.2   Partitioned Tablespace SMS

```
CREATE REGULAR TABLESPACE SOL#CADR IN NODEGROUP ALLNG
PAGESIZE 4096 MANAGED BY SYSTEM USING ('/db2/d001n003di/
e8412iol/e8412dba/SOL#CADR', '/db2/d002n003di/e8412iol/
e8412dba/SOL#CADR', '/db2/d003n003di/e8412iol/e8412dba/
SOL#CADR')ON NODES (3) USING ('/db2/d001n005di/e8412iol/
e8412dba/SOL#CADR', '/db2/d002n005di/e8412iol/e8412dba/
SOL#CADR', '/db2/d003n005di/e8412iol/e8412dba/SOL#CADR')ON
NODES (5)
EXTENTSIZE 32
PREFETCHSIZE 96
BUFFERPOOL ALLNG
OVERHEAD 24.100000
TRANSFERRATE 0.900000;
```

60.0.3   Partitioned DMS Table Space

```
CREATE REGULAR TABLESPACE  SHSDAU02
IN NODEGROUP NG0TO7
PAGESIZE  4096
MANAGED BY DATABASE
USING (FILE '/udb/db01/hsdprd01/hsddb01/tablespaces/hsdprd01/NODE0000/SQL00001/chau0201  15904') on nodes (0)
USING (FILE '/udb/db02/hsdprd01/hsddb01/tablespaces/hsdprd01/NODE0000/SQL00001/chau0202' 15904') on nodes (1)
USING (FILE '/udb/db03/hsdprd01/hsddb01/tablespaces/hsdprd01/NODE0000/SQL00001/chau0203' 15904') on nodes (2)
USING (FILE '/udb/db04/hsdprd01/hsddb01/tablespaces/hsdprd01/NODE0000/SQL00001/chau0204' 15904') on nodes (3)
USING (FILE '/udb/db05/hsdprd01/hsddb01/tablespaces/hsdprd01/NODE0000/SQL00001/chau0205' 15904') on nodes (4)
USING (FILE '/udb/db06/hsdprd01/hsddb01/tablespaces/hsdprd01/NODE0000/SQL00001/chau0206' 15904') on nodes (5)
USING (FILE '/udb/db07/hsdprd01/hsddb01/tablespaces/hsdprd01/NODE0000/SQL00001/chau0207' 15904') on nodes (6)
USING (FILE '/udb/db08/hsdprd01/hsddb01/tablespaces/hsdprd01/NODE0000/SQL00001/chau0208' 15904') on nodes (7)
EXTENTSIZE 32
PREFETCHSIZE 128
BUFFERPOOL BPLARGE
OVERHEAD 24.100000
TRANSFERRATE 0.900000;
```

60.0.4 Partitioned SMS Table Space

```
CREATE REGULAR TABLESPACE  SHSDAU01
IN NODEGROUP NG123
PAGESIZE  4096
MANAGED BY SYSTEM
USING ('/udb/db01/hsdprd01/hsddb01/tablespaces/hsdprd01/NODE0000/SQL00001/chau201') on nodes (1)
USING ('/udb/db02/hsdprd01/hsddb01/tablespaces/hsdprd01/NODE0000/SQL00001/chau202') on nodes (2)
USING ('/udb/db03/hsdprd01/hsddb01/tablespaces/hsdprd01/NODE0000/SQL00001/chau203') on nodes (3)
EXTENTSIZE 32
PREFETCHSIZE 128
BUFFERPOOL BPLARGE
OVERHEAD 24.100000
TRANSFERRATE 0.900000;
```

61.0   Agents and Threads For future use

62.0 Compression for future use

63.0   Connection Pooling

The Connection Pool maintains details of all connections in order to save having to re-establish connections again and again whenever it is needed to be used. Once a connection has been created and placed in a pool, an application can reuse that connection without performing a complete connection process. The connection is pooled when the application disconnects from the ODBC data source, and will, later, be given a new connection whose attributes are the same as the connection the application previously relinquished.

ODBC connections in MTS (Microsoft Transaction Server) COM objects have connection pooling turned on automatically whether or not the COM object is transactional. For multiple MTS COM objects participation in the same transaction the connection can be re-used between two or more COM objects in the following manner
which may cause a problem:-

Let us say there are two COM objects COM1 and COM2 that connect to the same ODBC data source and participate in the same transaction. After COM1 connects and completes it's work it then disconnects and the connection joins the connection pool. However this connection is reserved for use by other COM objects of the same transaction. It will be available to other transactions only after the current transaction ends.

When COM2 is invoked in the same transaction, it is given the recently pooled connection released by COM1. MTS ensures that the connection can only be given to COM objects participating in the same transaction.

On the other hand, if COM1 does not disconnect, it will tie up the connection until the transaction ends. When COM2 is invoked, in the same transaction, a separate connection will be acquired. Subsequently, this transaction ties up two connections instead of one.

This will result in timeouts, deadlocks and wait states. It could also affect WebSphere applications.

This reuse of the connection feature for COM objects participating in the same transaction is preferable as follows:-

- It uses fewer resources in both the client and the server. Only one connection is needed.

- It eliminates the possibility that two connections participating in the same transaction (accessing the same database server and accessing the same data) can lock one another, because DB2 servers treat different connections from MTS COM objects as separate transactions.

- Alternatively, the environment can be patched using an IBM Fix Pack. There are two fix packs from IBM available. Fix Pack 4 & 5 and Fix Pack 6. Fix Pack 6 contains the eFix which is the preferred method.

64.0 Unix for future use

65.0 Speeding up Performance revisited

For a large-scale cluster of WebSphere Application Servers, each of the WebSphere servers could install a DB2 Runtime Client (which includes the JDBC driver).
DB2's Java support includes support for JDBC, a vendor-neutral dynamic SQL interface that provides data access to your application through standardized Java methods. JDBC is similar to DB2 CLI in that you do not have to precompile or bind a JDBC program. As a vendor-neutral standard, JDBC applications offer increased portability – a required benefit in today's heterogeneous business infrastructure. An application written using JDBC uses only dynamic SQL. The JDBC interface imposes additional processing overhead for obvious reasons.

The new trends in application development have created a myriad of naming schemes and conventions that can often prove to be the most confusing part. Let's look at the basic information you need to know to make sense of it all: versions, drivers, and kits.

A common source of confusion with Java is the fact that there are different versions of JDBC and different types of JDBC drivers that can be used by different versions of Java. What's more, the versions of Java have their own nicknames and their own Java Development Kits (JDKs).

DB2 Universal Database, for the most part, supports JDBC Versions 1.2, 2.0, and 2.1 – with some operating system restrictions. Currently, work is under way for Sun Microsystems to complete the JDBC Version 3.0 specification which has just recently moved into its final draft.

Though not an official statement from within the DB2 development labs, there has not been any changes between JDBC Version 2.0 and JDBC Version 2.1 with respect to a DB2 environment, yet! DB2 Version 7.2 provides certifiable support for different versions of JDBC. However, you can still attain JDBC-specific certification for different versions and not fully implement the standard (minor exceptions are allowed).

Java Driver Types

A JDBC Type 1 driver is built into Java and basically provides a JDBC-ODBC bridge. DB2 Version 7 does support this type of driver since it is just an ODBC conversion; however, it is typically not used any more.

The DB2JDBC Type 2 driver is quite popular and is often referred to as the **app driver**. DB2 Version 7 supports an app driver. The app driver name comes from the notion that this driver will perform a native connect through a local DB2 client to a remote database, and from its package name (COM.ibm.db2.jdbc.app.*).
In other words, you have to have a DB2 client installed on the machine where the application that is making the JDBC calls/runs. The JDBC Type 2 driver is a combination of Java and

native code, and will therefore always yield better performance than a Java-only Type 3 or Type 4 implementation.

66.0 IT Myths and Realities

Myth 1 – Logical design should always map exactly to the physical.

Reality:- Physical design should remain as close as possible to the logical structure but changes in the physical design mandated by performance should never be ignored just because they did not come from the logical.

Myth 2 – Put everything in one bufferpool and let DB2 manage it.

Reality:- While this method is much touted and even used in some shops, you really want to do this only if your memory is limited to 10,000 4K pages or less or you do not have time to manage bufferpools or you are not all concerned about performance. Otherwise, use multiple bufferpools and wisely!

Myth 3 – Applications should be coded to follow the logical process.

Reality:- Pseudo-code or a logical process diagram never takes into account coding methods for performance. This is most dramatic in OLTP transaction coding.

Myth 4 – Most processes cannot be performed in SQL

Reality:- Actually, the reverse is generally true. SQL is a very rich language that can perform most processes. The real difficulty is that SQL is approached as an I/O handler instead of a set processor.

Myth 5 – Code and Reference tables should be used with DB2 declared referential integrity (RI).

Reality:- RI should not be used as a shortcut for edit validations, which generally belong elsewhere, but should be used for true parent child-relationships.

Myth 6 – Tables should only have one or two indexes at most.

Reality:- Tables should have as many indexes as required to provide the performance necessary.

Myth 7 – Large tables should be split into smaller tables.

Reality:- This is a legacy fear that too much data in a table means performance degradation. With several tables in the industry of over 6 billion rows in production, this myth has been laid to rest.

Myth 8 – DB2 defaults are OK.

Reality:- DB2 defaults are generally never OK, since they can and do change over versions and releases.

67.0   Errors

When an error occurs for which there is additional internal information that would be useful in problem diagnosis, DB2 creates *dump files.* The information contained within the dump files is mostly IBM confidential information (such as internal control blocks and structures), so they are created in a binary format. The primary audience for the Dump Files is the DB2 Customer Service personnel, so they are not much use to users and administrators. However, when they are created in the DIAGPATH, administrators should know to look in the db2diag. log file to try and determine what may have caused the anomaly. The types of problems that could cause these files to be created vary quite a lot, but they are usually related to some sort of severe error, such as a bad page in the database.

Dump files are named following the format based on the platform on which they are created.

UNIX <pppppp>.<nnn>, where <pppppp> indicates the PID and <nnn> indicates the node where the problem occurred.

In keeping with best practices, you should always supply these files to DB2 Customer Service when you contact them for support because dump files are important for defect investigation.

Core Files

When certain problems are encountered on UNIX based operating systems, these files might be created in core dump directories under the instance home sqllib/db2dump directory (for example $HOME/sqllib/db2dump), not the DIAGPATH. These files are called *core files.* If a program terminates abnormally, a core file is created by the system which stores a memory image of the terminated process. The core files contain the data region stack, and attached shared segments for the process, as well as additional information about the process. Errors such as memory address violations, illegal instructions, bus errors and user generated quit signals cause core files to be dumped.
Note that System Core Files are distinct from DB2 Core Files.

DB2 core files are each located within a directory for the offending process. The naming convention of the directory is similar to the UNIX dump file directory naming convention but the file itself is simply named "core". The directory names start with the letter 'c' followed by the PID of the affected process and the three digit directory suffix of the failing node. For example, where $HOME indicates the home directory of the instance owner user ID, $HOME/ sqllib/db2dump/c71198.010 would be a directory containing a core file for a problematic DB2 process with PID 71198 running on node 10. At the same time you may discover that dump and locklist dump files were created in DIAGPATH or the same PID.

As of Version 4.3 of AIX, DB2 core dumps include shared memory segments, so they are much larger than they used to be, however, changing the use-pre-430 parameter to "true" will override this, causing only the data segment to be dumped.

68.0   Quick Reference Tuning for DB2 UDB EEE (ESE)

The Extended Enterprise Edition from the DB2 product family supports data partitioning across clusters of massively parallel computers. A partitioned database can maintain very large amounts of data and can open opportunities for new applications. You can install DB2 EEE with multiple partitions on a single system (logical nodes), or you can distribute it on multiple systems as physical nodes. The database manager on each node manages part of the database. DB2 EEE architecture can improve the performance of decision support and online transaction processing systems.

Shared-nothing architecture

DB2 EEE works as a shared-nothing architecture on the Windows, UNIX and Linux operating systems.

The portion of the database that consists of its own data, indexes, configuration files and transaction logs is referred to as a database partition.

DB2 EEE's shared-nothing architecture allows database partitions to be assigned to one or more processors. The database partition does not share any data, but communicates via messages.

Parallelism

DB2 EEE supports two types of parallelism: inter-partition and intra-partition.

With inter-partition parallelism, for any given query, DB2 first identifies the partitions where the data resides. Then, the node from which the query originates (the coordinator node) coordinates with the other partition nodes to find all qualifying rows. Each node works on only a part of the data.

With intra-partition parallelism, the given query is divided into a series of operators such as scanning, joining and sorting but all the work for those operators is done concurrently in the same partition using different processes.

DB2 EEE process/thread model per partition

A DB2 EEE instance on Windows starts as a service when the db2start command is issued. Starting an instance starts db2syscs.exe for every partition on the node. These processes communicate via messages using IPC or TCP/IP. Each application is assigned an agent that works on behalf of the application.

Activating a database starts the logger for logging activity and the deadlock detector for detecting the deadlocks. DB2's license manager starts as a different process and monitors the license usage.

How DB2 uses memory

DB2 memory gets allocated at different stages.

Global control block

Memory space required for the database manager to run. This memory is allocated when the database manager starts. Fast Communication Manager (FCM) provides the communication support for DB2 EEE. FCM buffers are allocated from this memory area.

Database global memory

Memory required for a database to run. This memory is allocated when the database gets activated. The number of memory segments is limited by the *numdb* (number of databases) configuration parameter. The total size of database global memory is determined by the database configuration parameters.

Parameter Description

buffpage	Impacts the buffer pools whose size is set to -1
pckcachesz	Package cache size
util_heap_sz	Utility heap size
dbheap	Database heap
locklist	Maximum storage for locks

Application global memory

Application global memory is the memory required for the application to run. This memory is allocated when the first agent to receives a request from the application that requests a connection. This memory is controlled by the database configuration parameter app_ctl_heap_sz. Application global memory is shared between all agents working for the same application.

Agent private memory

Each agent has its own private memory. The lower of the configuration parameters maxappls and maxagents limits the number of memory segments:

Parameter Descriptions

maxappls	The total number of maximum applications for all active databases.
maxagents	The maximum number of agents.
agent_stack_sz	Agent stack size
udf_mem_sz	UDF shared memory set size
applheapsz	Application heap size
sortheap	Sort heap size
stmtheap	Statement heap size
stat_heap_sz	Statistics heap size
query_heap_sz	Query heap size
drda_heap_sz	DRDA heap size

Agent/application shared memory

This memory is shared between the agents that are working for the same application.
Agent shared memory is affected by the database and database manager configuration
parameters.

Parameter Descriptions

aslheapsz	Application support layer heap size
rqrioblk	Client I/O block size
applheapsz	Application heap size
agent_	stack_sz Agent stack size
stat_heap_sz	Statistics heap size
udf_mem_sz	UDF memory
sortheap	Sort heap size
rqrioblk	Query heap size
query_heap_sz	Client I/O block
drda_heap_sz	DRDA heap size

Creating the database

In DB2 EEE, issuing a simple command from the DB2 Command Line Processor (CLP)
prompt creates the database. The following example creates a database named *mydb*.

CREATE DATABASE *mydb*

After the above statement is executed, the database *mydb* appears in the instance directory.

Tablespaces

A tablespace is a logical layer between the database and the container objects that actually
hold the data. DB2 EEE supports two types of tablespaces: system-managed space (SMS)
and database-managed space (DMS). In SMS, the operating system's file manager allocates
and manages the storage space. The containers used in SMS must be directories. In DMS,
the database manager controls the storage space. The containers used in DMS must be a pre-
allocated file or a physical device, such as a hard disk drive. DBAs have the option to choose
the type of the tablespaces required for the catalog, and for temporary and user tablespaces.

If you use the defaults when creating a database, the following system-managed tablespaces
are created:

System - To store catalog information in system tables. The catalog contains information about
the definitions of the database objects (for example, tables, views, indexes, and packages),
and security information about the type of access that users have to these objects.
Temporary - To store the system temporary tables created during database processing.
User - To store the database objects like tables and indexes created by a user.
The following example illustrates how database administrators can create a database using
both DMS and SMS tablespaces. The following example creates the database named *tested*

with a system-managed catalog tablespace and database-managed temporary and user tablespaces:-

```
CREATE DATABASE tested
CATALOG TABLESPACE
PAGESIZE 4096
MANAGED BY DATABASE USING
(FILE 'C:\DATA\CAT1.DAT' 3000,
FILE 'D:\DATA\CAT1.DAT' 3000)
EXTENTSIZE 12
PREFETCHSIZE 16
TEMPORARY TABLESPACE
MANAGED BY SYSTEM USING
('C:\DATA\TEMP1' , 'D:\DATA\TEMP2')
USER TABLESPACE
MANAGED BY DATABASE USING
('\\.\PhysicalDrive1' 4096,
'\\.\PhysicalDrive2' 4096)
```

<u>PAGES</u>: Rows of table data are organized into blocks called pages. DB2 supports 4 KB, 8 KB, 16 KB and 32 KB page sizes. You can choose different page sizes for the different tablespaces based on the type of the application that is used. By default DB2 uses the 4 KB pages.

PAGESIZE: A 4 KB-page size is specified for the catalog tablespace.

EXTENT: A set of pages grouped into allocation units called extents. In the above example, EXTENTSIZE defines 12 4 KB pages to be grouped together for an extent.

PREFETCH: It is a method that allows DB2 to read the data in advance to prevent wait times while the data is being retrieved.

PREFETCHSIZE: Defines the number of pages that can be prefetched.

The command LIST TABLESPACES SHOW DETAILS displays the above information and more, including the total pages, usable pages, used pages, free pages, and so on.

After you find out a tablespace ID from the LIST TABLESPACES command, you can find more information about the containers by issuing the command LIST TABLESPACE CONTAINERS FOR 2 SHOW DETAIL. Where 2 is the ID of the tablespace. This command lists the containers for the specified tablespace. For every container it displays:- ID, name, type, total pages, useable pages, and whether the container is currently accessible or not.

With DB2, you can put index, data in separate tablespaces. Using different hard disk drives for the containers reduces the I/O contention. The following example creates the table *employee*. Regular data will be placed in tablespace *TBS1* and index data will be placed in *TBS2*.

```
CREATE TABLE employee (
E_NO INT NOT NULL,
E_NAME CHAR(20) NOT NULL )
IN TBS1 INDEX IN TBS2
```

Creating the buffer pools

DB2 uses buffer pools to cache the table and index data as they are being read or written to the hard disk drive. When a tablespace is created, it is assigned to the default buffer pool IBMDEFAULTBP. The DBA can map the buffer pool to the tablespace when creating or altering the tablespace. For every different page size tablespace, there should be a matching page size buffer pool.

```
CREATE BUFFERPOOL bp1
SIZE 4000
PAGESIZE 4096
```

The SQL statement above creates a buffer pool *bp1* with 4000 4KB pages (4000*4K = 16000K).

The existing tablespace can now be altered to map to buffer pool *bp1*. The following example maps the *emp* tablespace to buffer pool *bp1*:

```
ALTER TABLESPACE emp BUFFERPOOL bp1
```

Buffer pools are allocated as a number of shared memory segments on the database server. Database records are read and updated in the buffer pool area of memory. All buffer pools are allocated when the database is activated. As an application requests data out of the database for a specific table or index, pages containing that data are transferred to the associated buffer pool from the hard disk. Pages are not written back to hard disk until one of the following occurs:

All the applications are disconnected from the database
A new page needs to be read into the buffer pool
A page cleaner is available and is activated by the database manager

The buffer pool should be large enough to keep the required data in memory so that hard disk drive activity can be reduced. To take advantage of the increased buffer pool, you must rebind the applications for the optimizer to consider the buffer pool size when deciding its access strategy.

Configuring the database instance

A DB2 EEE instance or the instance manager resides on the catalog node. The instance owning machine (that is, node 0) owns the shared directory where this information is stored. Other database-partitioned servers that are added to an instance are said to be participating in the instance.

Physical nodes

You can configure physical nodes by installing the instance-owning database-partitioned server on one machine, and then installing a database-partitioned server on each of the other machines that participates in the partitioned database system.

Logical nodes

Having more than one database-partitioned server on the same machine is known as a multiple logical nodes (MLN) configuration. MLN takes advantage of SMP architecture. Use the db2ncrt command to add database-partitioned server nodes (a logical node) on your instance to create an MLN configuration.

The following example adds a new database-partitioned server to the instance *BENCH* on the instance owning machine node1. The command sets this new node to node1 using logical port 1.

db2ncrt /n:1 /u:*BENCH*\db2inst1,bmdb2 /I:*BENCH* /m:node1 /p:1 /h:node1

Nodegroups

A named subset of one or more database partitions is called a nodegroup. The subset that consists of more than one database partition is known as a multi-partition nodegroup. Multi-partition nodegroups can exist only within the database partitions that belong to same database.

The following example creates the nodegroup *allnodes* on all the database partitions:

CREATE NODEGROUP *allnodes* ON ALL NODES

This example creates the nodegroup *snodes* only on nodes 0 to 3 and node 5:

CREATE NODEGROUP *snodes* ON NODE (0 TO 3, 5)

After you create nodegroups, you must associate them with tablespaces to specify the nodes where the data is going to be partitioned.

```
CREATE TABLESPACE org
IN NODEGROUP snodes
MANAGED BY DATABASE
USING (DEVICE '\\.\PhysicalDrive1' 2000)   ON NODE(0)
USING (DEVICE '\\.\PhysicalDrive1' 2000)   ON NODE(1)
USING (DEVICE '\\.\PhysicalDrive1' 2000)   ON NODE(5)
```

69.0 Tuning guidelines

DB2 includes tools for monitoring and analyzing database performance problems.

Explain
Explain is very useful for identifying performance issues. Explain helps determine whether additional indexes are required, a query needs to be rewritten, locking strategies are appropriate, etc. Explain snapshot information can be captured in the following ways.

Enabling the EXPLAIN SNAPSHOT special register

Set the special register CURRENT EXPLAIN SNAPSHOT to YES to capture the snapshot for dynamic SQL statements. You can imbed the following statement in an application program or issue it interactively. By setting the special register, an Explain snapshot will be taken for any subsequent eligible dynamic SQL statements.

SET CURRENT EXPLAIN SNAPSHOT YES

Collecting explain snapshot on SQL procedures

The stored procedure needs to set the EXPLSNAP register to ALL or YES in order to collect the access plans.

db2 PREP EXPLSNAP {YES / ALL / NO }
            YES = static SQL
            ALL = static and dynamic
            NO = No snapshot

When the package is prepared with EXPLSNAP YES or ALL, you can obtain the plan for the entire package.

The db2expln tool describes the access plan selected for static SQL statements in the package stored in the system catalog tables, and the dynexpln tool describes the access plan for dynamic SQL statements.

The following example uses db2expln to retrieve the access plan for the package named *neword* in the database *mydb*, created by user *myuser* to a output file *output.file*.

db2expln -d mydb -p neword -c myuser -o output.file

70.0 Utilities that affect performance

DB2 provides various utilities, such as RUNSTATS, REORG and REORGCHK, to improve database performance.

The RUNSTATS utility updates the statistics in the system catalog tables to help with the query optimization process. With these statistics, the information database manager could make a decision that would increase the performance of an SQL statement. Use this utility after massive changes to the data and possibly after running REORG.

REORG eliminates the fragmentation in tables and indexes and may optionally order the rows of the table according to the order of the index. Use REORG when another utility, REORGCHK , indicates that REORG is needed and when the performance degrades over time - when data inserts, updates and deletes cause the clustering or space utilization to degrade.

REORGCHK examines the data in the system tables and applies formula to determine whether to reorganize of the table and its indexes. REORGCHK can also invoke RUNSTATS before examining the statistics. Run REORGCHK periodically, or when users notice degraded performance.

Examples:-

The following RUNSTATS example collects all possible statistics based on the indexes:

RUNSTATS ON TABLE *bench.neword* WITH DISTRIBUTION AND DETAILED INDEXES ALL

The following REORG example reorganizes the customer table using the system temporary tablespace TEMPSAPCE1 as a work area to store the intermediate results.

REORG TABLE *bench.customer* USING TEMPSPACE1

The following REORGCHK command examines whether the REORG is required and also updates the statistics on table *bench.customer*:

REORGCHK UPDATE STATISTICS ON TABLE *bench.customer*

Minimize I/O
Reduce I/O by taking the advantage of indexes, caching, reduced logging overhead, summary tables, and reduced fetches.

Use indexes

Creating proper indexes improves the performance of the query. DB2 index adviser (DB2ADVIS) provides assistance in the designing of indexes on tables. It is useful in the following situations:

Finding the best indexes for a problem query.
Finding the best indexes for a set of queries subject to resource limits that optimally applied.
Testing out an index on a workload without having to create the index.
Creating indexes helps avoid unnecessary table scans and sorts, speeds up frequently executed queries, and ensures uniqueness.

After you create indexes, use Explain to ensure that DB2 is using them.

Catalog cache, package cache, and the log buffer

Set the catalog cache, package cache, and log buffer size to an appropriate size to improve performance. The catalog cache is used to store table descriptor information that is used when the tables, views, or alias have been referenced in previous statements. The package cache sets the amount of database global memory to be used for caching a package's static and dynamic SQL statements. The log buffer holds log records in storage until they are written to a hard disk drive. Log records are written to a hard disk drive when one of the following occurs:

A log buffer is full.
A transaction or a group of transactions commits.

It is important that the log buffer be able to hold the amount of log space used by an average transaction. If not, you will experience poor logging performance due to a high number of log writes, mainly caused by the log buffer's "full condition".

Here are the commands you use to change these caching parameters:

```
db2 update db cfg for mydb using logbufsz 4096
db2 update db cfg for mydb using catalogcache_sz 1024
db2 update db cfg for mydb using pckacchesz 4096
```

Improving I/O parallelism The following are guidelines for improving I/O parallelism:-

Spread the data across multiple hard disk drives to reduce I/O wait time.
Separate data and indexes to different tablespaces to improve the performance of OLTP and OLAP applications by reducing I/O contention.
As a rule of thumb for OLTP applications, use DMS tablespaces with multiple devices.
Temporary and catalog tablespaces must be placed SMS.

Placement of log files

For OLTP applications, it is very important to place the log files on a separate physical device where their high activity will not negatively impact other work on the same hard disk drive.

DB2 parallel I/O

When reading data from or writing data to tablespace containers, DB2 may use parallel I/O if the number of containers in the database is greater than one. However, there are situations when it would be beneficial to have parallel I/O enabled for single container tablespaces, especially striped RAID devices.

Enabling intra-partition parallelism
To enable the optimizer to choose the intra-partition parallelism you must set the intra_ parallel database manager configuration parameter as shown here:

    db2 update dbm cfg using intra_parallel YES

In conjunction to intra_prallel one needs to set the dft_degree database parameter to specify the degree of the parallelism. This value should be set to greater than 1 to maximum to the number of processors that database partition can use.

The dft_degree database configuration parameter specifies the default level of parallelism for each database. A value of 1 means no intra-parallelism. A value of -1 means the optimizer determines the degree of parallelism based on the number of processors and the type of query.

Inter-partition parallelism gets enabled automatically when the database is partitioned across the nodes.

Summary tables

A summary table is a table whose definition is based on a result of a query. As such, the summary table typically contains pre-computed results based on the data existing in the table or tables that its definition is based on. If the SQL compiler determines that a dynamic query will run more efficiently against summary table than the base table, the query executes against the summary table, and you obtain the results faster than you otherwise would.

71.0   How data is partitioned

A partitioned key is a column, or group of columns, used to determine the partition in which the particular row of data is stored. You define a partition key for a table on the CREATE TABLE statement. If you do not specify a partitioning key for a table in a tablespace that is divided across more than one database partition in a nodegroup, DB2 creates one by default from the first column of the primary key. If no primary key is specified, the default-partitioning key is the first non-long field column defined on the table.

```
CREATE TABLE customer (
C_ID INTEGER NOT NULL,
C_FIRST VARCHAR(20) NOT NULL,
C_MIDDLE VARCHAR(20) NOT NULL)
IN CUSTOMER
INDEX IN CUSTOMER_IDX
PARTITIONING KEY(C_ID) USING HASHING;
```

Tuning using database snapshot monitors
Database monitors provide large amount of the information to the DBAs about database and the application accessing the database. Different monitor group switches are listed below:-

SORT   -   Amount of heap used, overflows and sorts performed
LOCK   -   Number of locks held, deadlocks and lock waits escalations
TABLE   -   Measures of activity (rows read, rows written)
BUFFERPOOL -   Physical reads, logical reads and timing information
STATEMENT   -   Number of commits, rollbacks, selects and failures

You can use these switches with the following commands to operate the snapshot monitor:

```
GET MONITOR SWITCHES
UPDATE MONITOR SWITCHES USING <switch-name> ON/ OFF
RESET MONITOR ALL/ FOR DATABASE <dbname>
```

71.0.1   Tuning buffer pools

The buffer pool hit ratio indicates the percentage of time that the database manager did not need to load a page from hard disk drive in order to service a page request; that is, the page was already in the buffer pool. The greater the buffer pool hit ratio, the lower the frequency of hard disk drive I/O. The following formula calculates the buffer pool hit ratio from the buffer pool snapshot information:

```
(1- ((pool_data_p_reads + pool_index_p_reads) /
(pool_data_l_reads + pool_index_l_reads))) * 100
```

If the hit ratio is low, increasing the number of buffer pool pages may improve performance.

71.0.2 Tuning the sort heap

The sort average elapsed time from the snapshot data can be measured as total_sort_time / total_sorts. As sort performance improves, this average will decrease. Increasing the sort heap may eliminate the merge phases, hence the improvement in sort time.

71.0.3 Tuning the sort heap threshold (sortheapthres)

Always keep the threshold high enough to fulfill the total sort heap required by all the applications, if not, the application request for a piped sort will be rejected. The percentage of piped sort requests that have been accepted by the database manager may be calculated by using the following formula:

$$100\% * piped\_sorts\_accepted / piped\_sorts\_requested$$

A small percentage is the indication that the performance can be improved by increasing the *sortheapthres* parameter.

71.0.4 Tuning the package cache (*pckcachesz*)

The package cache hit ratio can be calculated using the following formula:

$$(1 - (Package\ cache\ inserts\ /\ package\ cache\ lookups)) * 100$$

A smaller hit ratio indicates that the *pckcachesz* parameter should be increased.

71.0.5 Tuning the catalog cache

Catalog cache hit ratio can be calculated by using the following formula:

$$(1 - (Catalog\ cache\ inserts\ /\ catalog\ cache\ lookups)) * 100$$

A smaller hit ratio is the indication that the catalogcache_sz should be increased.

71.0.6 Tuning the max # of concurrent applications

The database configuration parameter maxappls specifies the maximum number of concurrent applications that can be connected to a database. Increasing the value of this parameter without lowering the maxlocks parameter or increasing the locklist parameter could cause you to reach the database limit on locks (locklist) rather than the application limit and as a result cause pervasive lock escalation problems. You should ensure that there are enough agents are available (maxagents) to the applications.

71.0.7 Tuning the maximum number of database manager agents (maxagents)

The maxagents indicates the maximum number of database manager agents available at any given time to accept application requests. The value of the maxagents parameter should be the sum of the values for maxappls in each database allowed to be accessed concurrently.

71.0.8 Fast Communications Manager (FCM)

DB2 can use FCM to communicate between agents working on the same request. The configurable parameters for the FCM are fcm_num_anchors, fcm_num_buffers, fcm_num_connect, and fcm_num_rqb. As the number of partitions goes up, the number of connection entries (fcm_num_connect) and BQS RQB (fcm_num_rqb) will go up. As the number of request blocks goes up, the number of FCM buffers (fcm_num_buffers) goes up. As the number of subsections (that is, the complexity of the query) goes up, the number of BDS RQB (fcm_num_rqb) and message anchors (fcm_num_anchors) will go up.

In an MLN environment, set DB2_FORCE_FCM_BP to YES, which allows DB2 to create FCM buffers in a separate memory segment. When the FCM buffers are created in a separate memory segment, the communication between FCM daemons of different logical partitions on the same physical node occurs through shared memory.

Disable file system caching for Windows
File system caching is performed as follows:-

For DMS file containers (and all SMS containers), the operating system may cache pages in the file system cache.
For DMS device container tablespaces, the operating system does not cache pages in the file system cache.
When working on Windows NT, the registry variable DB2NTNOCACHE specifies whether or not DB2 will open database files with a NOCACHE option. If DB2NTNOCACHE=ON, file system caching is eliminated. If DB2NTNOCACHE=OFF, the operating system caches DB2 files. This applies to all data except for files that contain LONG FIELDS or LOBS. Eliminating system caching allows more memory to be available to the database so that the buffer pool or sort heap can be increased. To disable file system caching, you must set the DB2NTNOCACHE registry variable to ON:-

    db2set DB2NTNOCACHE=ON

Eliminating system caching allows more memory to be available to the database so that the buffer pool or sort heap can be increased.

72.0 To execute a file with SQL Code in Telnet

db2   –tvf   in-filename   >   out-filename

Or make a ksh file:-

#! /usr/bin/ksh
db2 connect
db2 timestamp
db2 count
db2 "sql here"
db2 count
db2 timestamp
db2 terminate

73.0   FENCED AND NOT FENCED

The FENCED (opposite = NOT FENCED) option specifies that your function will always run in an address space that is separate from the database. This option causes a performance penalty due to process switching when the function is called, but it protects the integrity of the database against accidental or malicious damage that might be caused by the function. This clause is optional and the default is FENCED. One use of this clause is in User Defined Functions.

An unfenced function runs in the same address space as the database and can damage the integrity of your data if your function "runs amok" in the address space. In order to create a NOT FENCED function you must possess either SYSADM or DBADM authority, or a database level authority called CREATE_NOT_FENCED, which can only be granted by someone with SYSADM or DBADM authority. You can see who has been granted authority to create unfenced functions by looking in the NOFENCEAUTH column of the DBAUTH catalog table.

I strongly suggest that functions be run in FENCED mode until they are very well tested. You may declare a function FENCED mode at first and later convert it to NOT FENCED when it is working correctly. To do this you must drop the function and recreate it with a new CREATE FUNCTION statement specifying the NOT FENCED clause.

Note:- The performance penalty for using FENCED (the default) is dependent on the scope of the function in which it is used.

74.0 RAID DISK ARRAYS

In a RAID disk array, multiple disks are used and managed by a disk controller, complete with it's own CPU. All of the logic required to manage the disks forming this array is contained on the disk controller, therefore, this implementation is operating system independent.

Five types of RAID architecture exist, RAID 1 through RAIID 5, and each provides disk fault tolerance. Each varies in function and performance. In general, RAID refers to a redundant array. RAID-0, which provides only data striping (and not fault tolerant redundancy) is not addressed here. Although the RAID specification defines five architectures, only RAID 1 and RAID 5 are typically used today.

RAID 1 is also known as disk mirroring or duplexing. Disk mirroring duplicates data (at the file level) from one disk onto a second disk, using a single disk controller. Disk duplexing is the same as disk mirroring, except that disks are attached to a second disk controller (like two SCSI adapters). With this type of data protection, either disk can fail, and the data is still accessible from the other disk. With duplexing, a disk controller can also fail without compromising data protection. Performance is also good with RAID 1, but the trade-off in this implementation is that the required disk capacity is twice that of the actual amount of data because data is duplicated on pairs of drives.

RAID 5 involves data and parity striping by sectors, across all disks. Parity is interleaved with data information, rather than stored on a dedicated drive. Data protection is good, if any disk fails, the data can still be accessed by using the information from the other disks, along with the striped parity information. Read performance is good, although write performance is considerably worse than RAID 1 write performance. A RAID 5 configuration requires a minimum of three identical disks. The amount of extra disk space required for overhead varies with the number of disks in the array. In the case of a RAID 5 configuration of five disks, the space overhead is 20 percent.

When using a RAID disk array (not RAID 0), a failed disk will not prevent you from accessing data on the array. When hot pluggable or swappable disks are used in the array, a replacement disk can be swapped with the failed disk while the array is in use. With RAID 5, if two disks fail at exactly the same time, all the data is lost, but the probability of simultaneous disk failures is very small.

Consider RAID 1 disks for your logs because this provides recoverability to the point of failure, and offers good write performance, which is important for logs. In cases where reliability is critical and write performance is not so critical, consider using RAID 5 disks. Alternatively, if write performance is critical and you are willing to achieve this despite the cost of additional disk space, consider RAID 1 for your data as well as logs.

75.0   Identity columns

Identity columns let you automatically generate guaranteed unique sequential numbers for table rows. Gone are the days of SELECT MAX(ID)+10 to derive the next key number. Example 1 shows how to use an identity column for EMPNO.

When defining identity columns, you can choose from several options. The START WITH "n" clause lets you define the starting sequence number. If you don't specify the starting number, the default is "1." The INCREMENT BY "n" clause lets you define the increment; again, the default is "1." The CACHE "n" clause defines the quantity of sequence numbers. A table can have only one identity column; if you attempt to create two, you'll receive a SQLSTATE 428C1. The identity column must have a numeric data type and is implicitly NOT NULL. Note:- Identity Columns cannot be defined on a partitioned database.

The CACHE "n" clause stipulates that n number of incremental values will be stored in DBHEAP memory for quick assignment to rows as needed. The default value for n is 20, and n can range from 2 to 32,767. As sequence numbers are needed, they are popped off the unused stack in the cache. When the cache is exhausted, DB2 performs a synchronous I/O to the log to obtain the next set of sequence numbers for the cache. When the database is deactivated, any unused sequence numbers in the cache are discarded and can never be used again. Thus, database deactivation will cause forfeiture of sequence numbers and gaps in the numbers assigned to data rows. The NO CACHE clause causes a synchronous I/O to the log to occur for each and every sequential number assignment because no sequence numbers are pre-allocated. Using NO CACHE will be slower, but will prevent the sequence number gaps that can occur when the database is deactivated while CACHE is being used. If gaps in identity column values are acceptable (for example, you simply want to assign a unique employee number and aren't concerned whether the numbers are in sequence), the use of CACHE n is clearly the best performing choice. When setting CACHE size, keep it in proportion to your tolerance for forfeited numbers at each database deactivation and consider the data type, expected number of rows, and insert rate. If no caching option is specified, CACHE 20 is the default.

```
Example 1:- CREATE TABLE EMPLOYEE (
            EMPNO   INTEGER
            GENERATED ALWAYS AS IDENTITY
            (START WITH 1 INCREMENT BY 10 CACHE 10),
            FNAME   VARCHAR (30),
            LNAME   VARCHAR (30),
            ULNAME GENERATED ALWAYS AS (UCASE (LNAME)));
            CREATE UNIQUE INDEX EMPIX_U ON EMPLOYEE (EMPNO);
            CREATE INDEX EMPIX1 ON EMPLOYEE (ULNAME);
        INSERT INTO EMPLOYEE (FNAME, LNAME) VALUES ('James', 'Bond');
        INSERT INTO EMPLOYEE (FNAME, LNAME) VALUES ('Dean', 'Martin');
SELECT * FROM EMPLOYEE WHERE UCASE(LNAME) = 'BOND'
Or
SELECT * FROM EMPLOYEE WHERE ULNAME = 'BOND'
(The optimizer will use the EMPIX1 index for both select statements)
```

John V. McLean

```
EMPNO      FNAME      LNAME      ULNAME
-------    ------     -----      ------
   1    James   Bond  BOND
```

76.0   The Optimizer Revisited

Lower the filter factor of a query by adding redundant predicates. Example: Change SELECT LASTNAME FROM TOLADR WHERE WORKDEPT = 'IT'
To SELECT LAST NAME FROM TOLADR WHERE WORKDEPT = 'IT' AND WORKDEPT = 'SOFTWARE' AND WORKDEPT = 'DATA PROCESSING'

The redundant predicates are the last two 'where' statements and do not affect SQL statement functionality. However, DB2 calculates a lower filter factor, which increases the possibility that an index on the WORKDEPT column will be chosen. The lower filter factor also increases the possibility that the table will be chosen as the outer table, if the redundant predicates are used in a join (more about the filter factor later).

When redundant predicates are used to enhance performance, as detailed above, be sure to document the reason for the extra predicates. Failure to do so may cause somebody to assume they are an error and then remove them.

Another way to get a small performance boost from an SQL statement is to change the physical order of the predicates in the SQL code. DB2 evaluates predicates first by predicate type, then according to the order in which it encounters the predicates. The first four types are listed in the order DB2 processes them:- 1. Equality, 2. Ranges, 3. IN – where a column is tested against a list of values, 4. Place a predicate with fewer values in a table before a predicate with more values in the table, for example if there are more males in the entire company than workers in a specific department (which is very likely) then in the SQL statement:- SELECT LASTNAME FROM TOLADR WHERE WORKDEPT = 'DATATEAM' AND SEX = 'M'. Since there are less workers in the datateam than males in company, test the number of workers in the datateam first. This will shave a bit off the query's processing time.

77.0   REORGS

It is good to have an understanding of what goes on in a REORG, this way you can determine with a reasonable assurance what makes one REORG take much longer than another, even though at 'first glance' both tables or groups of tables being REORGed seem very similar.

<u>The REORG switch phase</u>

Before Version 7, online REORG tried to fully preserve the underlying dataset names for tablespaces and index spaces via a series of AMS RENAMEs that took place during the REORG switch phase. However, this process can cause significant contention in the system because the duration of the switch phase (during which the object being reorganized cannot be accessed) is directly proportional to the number of underlying datasets. At least one dataset exists for each tablespace and each index. So, if a multitable tablespace contains 100 tables and each table has an average of two indexes, there are typically around 300 data sets.

With V7 and up, you can avoid dataset renames in the switch phase. The fifth data set qualifier (called the instance node) changes from a fixed value of I0001 to a variable value, alternating between I0001 and J0001. In the "utility" phase of the REORG, if the original object's instance node starts with an "I", the shadow object's instance node starts with a "J" and vice-versa. In the switch phase, the catalog is updated and the shadow object becomes the active object. In the "utility" termination phase, the original object is deleted.

Specifying the FASTSWITCH option on the REORG statement activates the new no-rename behavior in the REORG switch phase.

<u>REORG timeout and drain retry options</u>

Before V7, the timeout value was at least equal to the value specified as the transactions timeout. It was calculated as a product of the system parameters IRLMRWT (the transactions timeout value) and UTIMOUT, the utility timeout factor (greater or equal to 1) that allows REORG to wait longer for a resource than a regular statement would. This behavior is not what most customers want. It's often more appropriate if REORG times out before any user transaction. In V7, the new REORG option DRAIN_WAIT (with the possible values of 1 to 1800 seconds) specifies how long a REORG will wait to acquire drains. Draining is a mechanism used to take over an object and serialize access to it. The new option applies to drains in the UNLOAD, LOG and SWITCH phases of the REORG. When multiple objects need draining (for example a tablespace and it's index) the DRAIN_WAIT value is the aggregate time. In contrast, the timeout value is as per V6 for other objects. Use the DRAIN_ WAIT option in conjunction with the new retry logic that Preserves REORG processing done before DRAIN_WAIT is exhausted, otherwise, after a timeout the REORG process would terminate without accomplishing anything.

78.0 TIMESTAMP SWITCH

For those who still remember assembler, here is routine for switching a timestamp around. Having a timestamp as a key with the microseconds leftmost saves reorg time, and speeds up inserts and retrieval (selects).

FUNCTION:- READ A TIMESTAMP (26 BYTES) FROM A CALLING PROGRAM AND RETURN IT AS AN INVERTED TIMESTAMP OF 18 BYTES TO THE CALLING PROGRAM. THIS ROUTINE USES THE SAME AREA OF THE ORIGINAL TIMESTAMP TO PLACE THE INVERTED TIMESTAMP. THE LAST 8 BYTES OF THE INVERTED TIMESTAMP ARE TRUNCATED.

```
TSCOVER   CSECT                    CONTROL SECTION NAME PRINT NOGEN
          USING *, 3               SET ADDRESSABILITY
          SAVE  (14, 12)           SAVE CALLING PROGRAMS REGS
          LR    3, 15              SET REG 3 AS THE BASE REG
          LA    2, SAVEA           SAVE AREA ADDRESS TO REG 2
          ST    13, 4(2)           SET CALLING PROGRAMS FWD POINTER
          ST    2, 8(13)           SET THIS ROUTINES BACKWARD POINTER
          LR    13, 2              MOVE ADDRESS OF CALLER SAVE AREA
          B     GOMAIN             BRANCH TO THE MAIN ROUTINE
SAVEA     DS    18F                SAVE AREAA FOR CALLING PROGRAMS
GOMAIN    L     6, 0(1)            GET ADDR OF TIMESTAMP FROM CALLER
          MVC   AREA, 0(6)     INITIALIZE DATA AREA WITH PASSED TS
          TR    INPTMS,CLEARI  TRANSLATE TS VALUES (INVERT)
          MVC   0(20,6), INPTMS    INVERTED TS TO RETURN AREA
          MVC   INPTMS,CLEARI      CLEAR INPUT ADDRESS AREA
          L     13,4(,13)          RESTORE CALLERS REGISTERS
          RETURN (14,12),RC=0      RETJURN TO CALLER
          LTORG
AREA      DC    CL26 ' '           LENGTH OF TS AREA
INPTMS    DC    X'19181716151412110F0E0C0B09080605030201000000000000'
CLEARI    DC    X'19181716151412110F0E0C0B09080605030201000000000000'
          END   TSCONVER
```

79.0 Will DB2 become invisible soon??

Here are some of the new features that are leading DB2 into invisibility:-

Block transactions on log directory full: Before V7 DB2 would crash if the log directory was full and a new log could not be created. In V7, DB2 will attempt to create the log every five minutes, while reporting the condition to the db2diag.log. This feature is controlled by a new registry variable called DB2_BLOCK_ON_LOG_DISK_FULL.

This new feature will allow read-only applications to continue, provided they are not dependent on rows locked by an update transaction. Programs and applications that are updating data will not complete their operation until the log has been created, in fact, they go into a long wait state.

Suspended I/O and split mirror backup: To meet the need for increased availability (now verging on 24 x 365.25), DB2 V7 includes new backup up and recovery options. Backup on very large databases requires substantial time and resources and can interfere with continuous availability. Many customers cannot afford off line or online backups on a 1TB database, cutting edge disk drive technology such as IBM's ESS (Enterprise Storage Server), RAMAC's Virtual Array Systems and EMC Corp's Clarion systems provide new opportunities for high availability through fault tolerance and redundancy.

The online split mirror function in DB2 V7 allows a consistent mirror of a database to split off while transaction processing continues on the live database with a limited performance hit. Recovery on the split copy is acceptable for consistency; backups and system copies can be done from a mirror image.

The db2inidb utility must be used on the split copy to indicate the intended use of the database. Use only logs from the primary database, not from the split image, and run db2inidb on all nodes on a EEE environment before the split image can be accessed.

This is the command flow:-

db2inidb <db-name> as snapshot
db2inidb <db-name> as standby
db2inidb <db-name> as mirror

Snapshot results in crash recovery, making the database consistent (a new log chain is started). You cannot roll forward through any of the logs from the original database.
The database will be fully available. Standby places the database in a roll forward pending state. Mirror will replace the original database.

80.0 Basic Access

To see the current values in the Database Configuration parameter:-

GET DATABASE CNFIGURATION

To update values for individual parameters in the database manager configuration file:-

UPDATE DATABASE CONFIGURATION USING <parameter name> <value>, <next parameter> <next value>, …..

To reset the values of all the configuration parameters:-

RESET DATABASE CONFIGURATION

<u>Some key parameters</u>

AVG_APPLS – Stands for average number of active applications. This parameter specifies the average number of active applications that are running against the database at any one time. The optimizer uses this parameter to estimate buffer pool usage when the application is run

BUFFPAGE – The default value for the Buffer Pool Size database configuration parameter on UNIX 32 bit operating systems is 1000 with a range from 2 to 524,288.
UNIX 64 bit operating systems also has a BPS of 1000 but with a range of 2 to 2 Giga Bytes.
The unit of measure is pages. To determine whether the buffpage parameter is active for a buffer pool use:-

SELECT * FROM SYSCAT.BUFERPOOLS

Each buffer pool showing NPAGES with a value of −1 uses the value of the BUFFPAGE parameter when creating the buffer pool.

CHNGPGS_THRESH – The default value for the Changed Pages Threshold database configuration parameter is 60% with a range of 5% to 99%. The % given is used to specify the threshold at which **a**synchronous **p**age **c**leaners get started. The **APC's** keep track of the pages of the buffer pool, and when the % is reached it writes the pages to disk. When changing this parameter consider the number of APCs identified by the NUM_IOCLEANERS parameter.

DFT_DEGREE – The default value for the Default database configuration parameter is 1 and the range is from 1 to 32,767 OR −1. This parameter specifies the default value for the CURRENT DEGREE special register and the DEGREE bind option. It is useful only if you are working in an *intrapartition parallelism environment.* This parameter is one of the factors used to determine whether SQL queries from applications will take advantage of multiple processors to handle queries. A value of −1 means that the optimizer determines the degree of parallelism based on the number of processors and the type of query requested.

LOCKKIST – The default value for the Maximum Storage for the Lock List database configuration parameter varies from 25 to 100 pages, based on the operating system and the environment. The value of this parameter in pages, indicates the amount of storage allocated to the lock list. There is only one Lock List per database, and it contains the locks held by all the applications currently connected to the database. Each lock takes up space in the Lock List, 72 bytes for the first lock on an object and 36 bytes for subsequent Locks on the same object. Lock escalation occurs when one application reaches the value of *maxlocks*.

Recommendation:- Perform frequent commits within your applications to release held locks.

LOGBUFSZ - The default value for the Log Buffer Size configuration parameter is 8 pages and the range is from 4 to 4096 or 65K depending on the operation system involved. Just as there is a buffer to hold data pages, there is also this buffer for the database logs. Every action against the database is recorded in a log record. The log buffer holds these records before writing them to disk. The value of this parameter must be equal or less than the *dbheap* configuration parameter.

MAXAPPLS –The default value for the Maximum Number of Active Applications database configuration parameter on UNIX based operating systems is 40, and on NT servers with local and remote clients it is 20. The range for all operating systems is from 1 tot 60,000. This parameter specifies the maximum number of concurrent applications
that can be connected to a database.

MAXLOCKS – The default value for the Maximum Percent of Lock List Before Escalation configuration parameter on UNIX platforms is 10 percent, and on all others it is 22 percent. The value of this parameter is the percentage of the Lock List that must be held by a single application before the database manager performs lock escalation for the locks held by an application.

MINCOMMIT – The default value for the Number of Commits to Group database configuration parameter is 1 and the range is from 1 to 25 commits that must be performed before log records in the log buffer are written to disk. Having the log records not written out for more than one commit reduces the database manager overhead associated with writing log records. Reducing overhead is significant to the overall performance of the database, especially when you have multiple applications with potentially many commits requested over a short period of time.

Recommendation: Increase the value of this parameter if multiple read/write applications Request concurrent database commits.

81.0 Myths and Reality in DB2 - to be continued

82.0   Errors

When an error occurs for which there is additional internal information that would be useful in problem diagnosis, DB2 creates *dump files.* The information contained within the dump files is mostly IBM confidential information (such as internal control blocks and structures), so they are created in a binary format. The primary audience for the Dump Files is the DB2 Customer Service personnel, so they are not much use to users and administrators. However, when they are created in the DIAGPATH, administrators should know to look in the db2diag. log file to try and determine what may have caused the anomaly. The types of problems that could cause these files to be created vary quite a lot, but they are usually related to some sort of severe error, such as a bad page in the database.

Dump files are named following the format based on the platform on which they are created.

UNIX <pppppp>.<nnn>, where <pppppp> indicates the PID and <nnn> indicates the node where the problem occurred.

In keeping with best practices, you should always supply these files to DB2 Customer Service when you contact them for support because dump files are important for defect investigation.

82.0.1 Core Files for future use

83.0   The Filter Factor for future use

84.0 Privileges

Grant and Revoke Examples

SET CURRENT SQLID = 'HDBT99';

GRANT BIND,EXECUTE ON PACKAGE YYY.* TO XXXXXX;

GRANT BIND,EXECUTE ON PLAN YYYYYYYY TO XXXXXX;

GRANT ALL ON YYY.* TO XXXXXX;

GRANT SELECT ON TABLE DB2T.ZZZZZZZZ TO XXXXXX;

GRANT SELECT ON TABLE DB2T.ZZZZZZZZ TO PUBLIC;

GRANT ALL ON TABLE DB2T.ZZZZZZZZ TO XXXXXX;

GRANT UPDATE ON TABLE DB2T.ZZZZZZZZ TO XXXXXX;

GRANT EXECUTE, BIND ON PACKAGE YYY.YYYYYYYY TO PUBLIC;

GRANT EXECUTE, BIND ON PACKAGE XXX.* TO AAAAAA; (WHERE XXX IS
THE COLLECTION ID).

GRANT SELECT
ON TABLE DB2P.ACDVBASE,
     DB2P.ACREBASE,
     DB2P.ACSSBASE,
     DB2P.ACTCBASE,
     DB2P.BBDPBASE,
     DB2P.BLOMBASE,
TO XXXXXX;

GRANT SELECT ON TABLE DB2T.ZZZZZZZZ TO PUBLIC;

GRANT SELECT ON TABLE DB2T.ZZZZZZZZ TO PUBLIC AT ALL LOCATIONS;

REVOKE ALL PRIVILEGES
ON TABLE DB2T.SHDABASE
FROM XXXXXX;

REVOKE ALL ON TABLE DB2P.ZZZZZZZZ
FROM XXXXXX;

```
REVOKE ALL ON TABLE DB2P.ZZZZZZZZ
  FROM PUBLIC;

SET CURRENT SQLID = 'HDBT99';
GRANT SELECT ON TABLE DB2T.VPEACCCM TO PUBLIC AT ALL LOCATIONS;
GRANT SELECT ON TABLE DB2T.VPEACCRR TO PUBLIC AT ALL LOCATIONS;
GRANT SELECT ON TABLE DB2T.VPEACCT  TO PUBLIC AT ALL LOCATIONS;
```

85.0 Drives for future use

86.0 OS/390 – Load, Image Copy, Recover and Runstats

```
//USERIDDD JOB (1,34),'DB2  LOAD ',CLASS=Q,MSGCLASS=H,
//    NOTIFY=&SYSUID
//***
//*** DELETE
//***
//S010   EXEC PGM=IEFBR14
//INDD01    DD DSN=DBAT.SYSDISC,SPACE=(TRK,0),
//       DISP=(MOD,DELETE,DELETE),UNIT=SYSDA
//***
//*** LOAD
//**
//S020    EXEC DSNUPROC,SYSTEM=HDBT,UID='LOADDB',UTPROC=''
//DSNUPROC.SORTWK01 DD UNIT=SYSDA,SPACE=(CYL,(200,200))
//DSNUPROC.SORTWK02 DD UNIT=SYSDA,SPACE=(CYL,(100,200))
//DSNUPROC.SORTWK03 DD UNIT=SYSDA,SPACE=(CYL,(100,200))
//DSNUPROC.SORTWK04 DD UNIT=SYSDA,SPACE=(CYL,(100,200))
//DSNUPROC.SYSERR   DD DSN=DBA024.SYSERR,UNIT=SYSDA,
//       DISP=SHR,SPACE=(CYL,(100,150))
//DSNUPROC.SYSMAP   DD DSN=DBA024.SYSMAP,UNIT=SYSDA,
//       DISP=SHR,SPACE=(CYL,(100,150))
//DSNUPROC.SYSUT1   DD DSN=DBA024.SYSUT1,UNIT=SYSDA,
//       DISP=SHR,SPACE=(CYL,(100,150))
//DSNUPROC.SORTOUT  DD UNIT=SYSDA,SPACE=(CYL,(100,150))
//DSNUPROC.SYSDISC  DD DSN=DBAT.SYSDISC,
//     DISP=(NEW,CATLG,DELETE),UNIT=SYSDA,VOL=SER=TEST91,
//     SPACE=(TRK,(1,15),RLSE)
//SYSREC00 DD DSN=DBA024.TEST1.DATA,DISP=SHR
//DSNUPROC.SYSIN  DD  DSN=DBAT.DBA024.PARMLIB(TEST1),DISP=SHR
//***
//*** DELETE THE COPY AND RECOVERY FILES
//***
//S030      EXEC PGM=IEFBR14
//INDD01    DD DSN=DBAT.DBA024I.IC.DBA024.TEST1,
//          DISP=(MOD,DELETE,DELETE),UNIT=SYSDA,
//          SPACE=(TRK,(1,1),RLSE)
//INDD02    DD DSN=DBAT.DBA024I.RC.DBA024.TEST1,
//          DISP=(MOD,DELETE,DELETE),UNIT=SYSDA,
//          SPACE=(TRK,(1,1),RLSE)
//***
//*** COPY AND RECOVER
//***
//*=============================================================
//S040      EXEC PGM=DSNUTILB,
//             PARM='HDBT,TEST1.COPY',
//             REGION=0M
//*********************************************************************
//* STEP DESCRIPTION: IMAGE COPY AND RECOVERY COPY TABLESPACE
//*********************************************************************
//* STEP RESTART INSTRUCTIONS:
//*     CONTACT THE ONCALL DBA; DO NOT UNCATALOG ANY DATASETS.
//*********************************************************************
//*
//STEPLIB DD DISP=SHR,DSN=DB2MVS.HDBT.DSNLOAD
//        DD DISP=SHR,DSN=DB2MVS.HDBT.DSNEXIT
//ABNLIGNR DD DUMMY
//SYSPRINT DD SYSOUT=*
//SYSUDUMP DD SYSOUT=*
//UTPRINT DD SYSOUT=*
//SYSOUT   DD SYSOUT=*
//SYSIN    DD *
  COPY TABLESPACE DBA024.TESTTS
  DSNUM ALL
```

```
      COPYDDN (SYSCOPY)
      RECOVERYDDN (RECOVRY)
      FULL YES  SHRLEVEL REFERENCE
//*
//*------------------------------------------------------------------
//*  UTILITY  RECOVERY SITE IMAGE COPY DD STATEMENTS
//*------------------------------------------------------------------
//RECOVRY  DD DSN=DBAT.DBA024I.RC.DBA024.TEST1,
//            DISP=(NEW,CATLG,CATLG),
//            UNIT=SYSDA,
//            SPACE=(CYL,(30,10),RLSE)
//*
//*------------------------------------------------------------------
//*  UTILITY COPY DD STATEMENTS
//*------------------------------------------------------------------
//SYSCOPY  DD DSN=DBAT.DBA024I.IC.DBA024.TEST1,
//            DISP=(NEW,CATLG,CATLG),
//            UNIT=SYSDA,
//            SPACE=(CYL,(30,10),RLSE),
//            DCB=(RECFM=FB,LRECL=4096,BLKSIZE=0)
//***
//*** EXECUTE RUNSTATS
//***
//*================================================================
//S050       EXEC PGM=DSNUTILB,
//              PARM='HDBT,TEST1.STATS',
//              REGION=0M
//**********************************************************************
//*  STEP DESCRIPTION: DB2 RUNSTATS
//**********************************************************************
//*  STEP RESTART INSTRUCTIONS:
//*      CONTACT THE ONCALL DBA; DO NOT UNCATALOG ANY DATASETS.
//**********************************************************************
//*
//STEPLIB  DD DISP=SHR,DSN=DB2MVS.HDBT.DSNLOAD
//         DD DISP=SHR,DSN=DB2MVS.HDBT.DSNEXIT
//ABNLIGNR DD DUMMY
//SYSPRINT DD SYSOUT=*
//SYSUDUMP DD SYSOUT=*
//UTPRINT  DD SYSOUT=*
//SYSOUT   DD SYSOUT=*
//SYSIN    DD *
  RUNSTATS TABLESPACE DBA024.TESTTS
     INDEX (ALL)
     SHRLEVEL REFERENCE
     REPORT NO  UPDATE ALL
     TABLE(DB2T.TEST1)
        COLUMN(ALL)
//*
//*================================================================
//
```

87.0 Log Archiving

With V7.2 and higher you can close and archive the active log at any time. For recoverable databases this feature lets you create a complete set of log files to a specific time for such purposes as off-site storage. To do so, use the following command:-

ARCHIVE LOG FOR DATABASE abcxyz USER name USING password

If other applications have transactions in progress, you'll notice a slight performance degradation as the ARCHIVE LOG command flushes the log buffer to disk. If user exits are enabled, an archive request is issued after the logs are closed and truncated. Completion of the archive command does not guarantee the logs have been moved to the archive directory, so you should to make sure.

Log Mirroring:- This high availability feature helps prevent log failures due to hardware failures, file system full conditions or accidental deletion of an active log. Hardware errors include such problems as channel corruption or device driver errors. The log mirroring feature supplements hardware disk mirroring as an additional write to the log.

Two new registry variables support log mirroring:- NEWLOGPATH and NEWLOGPATH2. These paths should be placed on separate physical devices so that if one of the devices fails, both copies will not be lost.

The db2set command, db2set NEWLOGPATH2=[1,ON,YES], specifies the secondary log path and concatenates a "2" to the path name. For example, if LOGPATH is /a/dbhome/sqllogdir/ logpath, then the secondary path for logging will be /a/dbhome/sqllogdir/logpath2.

For Unix systems, the file system can point to a separate device for the second copy. For Windows systems, the second log copy will be on the same device, which is not really desirable because a hardware failure could destroy both copies. Future releases will allow an explicit logpath name.

If DB2 encounters an error on either path, it won't use that logpath until the database attempts to access the next log file. The failing path will be marked and a message will be written to the remaining log path. If another failure occurs on the remaining good path the database will abend.

In earlier versions, DB2 would crash if the log directory was full and a new log file could not be created. In V7.2, DB2 will attempt to create the log every five minutes, while reporting the condition to the db2diag.log. This feature is controlled by a new registry variable called DB2_BLOCK_ON_LOG_DISK_FULL. This new feature will allow read-only applications to continue, provided they are not dependent on rows locked by an update transaction. Applications that are updating data will not complete their operation until the log has been created.

88.0   The Path to the Logs revisited

Configuration Parameters and the Path to the Logs

There are two levels of configuration parameters, one at the instance level (in this example initially connect to the instance exwintq007), and one at the database level, in this example connect to database xxxxxxxx.

To display the configuration parameters at the instance level (a.k.a. the database configuration <u>manager)</u> enter:-

db2 get dbm cfg >> dbcfgmgr.out

Direct the output to an output file which you may browse later.

To display the configuration parameters at the database level (a.k.a. database configuration) enter:-

db2 get db cfg for xxxxxxxx >> dbcfg.out

Direct the output to an output file which you may browse later. Note the database name is xxxxxxxx.

If you want to see the path to the logs and the log files, look for "Path to log files" as the parameter prompt, then change directories to the path to the logs. This information is contained in the database configuration parameters. The first log file is at S0012825.LOG. As you can see permission was denied in this instance.

<u>From the database configuration parameters for xxxxxxxxl:-</u>

```
Log file size (4KB)              (LOGFILSIZ) = 2500
Number of primary log files    (LOGPRIMARY) = 10
Number of secondary log files   (LOGSECOND) = 20
Changed path to log files      (NEWLOGPATH) = Path to log files

               =    /d001n001log/udbqol01/e8412qol/NODE0001/
First active log file
               = S0012825.LOG
```

Note:- In a partitioned database the NODE0001 is given as the default, other Node
numbers as appropriate may be used for each partition.

Attempt to access Path to Log Files in QA is denied – database e8412qol:-

```
$ pwd
/home/e8412dba
$ cd ..
$ cd ..
$ pwd
/
$
$ cd /prd/d001n001log/udbqol01/e8412qol/NODE0001/
ksh: /prd/d001n001log/udbqol01/e8412qol/NODE0001/: Permission
denied.
```

89.0 Sample OS/390 JCL for the DB2 Performance Monitor

```
//G2DGPM2R JOB (G,YV1,YV15000,4,B227),'JOHN MCLEAN X-4714',
//**G4DEGP14 JOB (G,YV1,YV15000,4,B227),'FRANK SLEDGE X-6875',
//****     CLASS=C,MSGLEVEL=(1,1),
//         CLASS=K,MSGLEVEL=(1,1),TIME=1440,
//         MSGCLASS=T,NOTIFY=G4DE
/*JOBPARM LINES=99999,ROOM=B227
//STEP01  EXEC PGM=DB2PM,REGION=0M
//STEPLIB  DD  DSN=TTAP.TS2.DB2PM.ISPLLIB,DISP=SHR
//INPUTDD  DD  DSN=PSD.K02.K2D8000.TH8.G4674V00,DISP=SHR,
//**INPUTDD  DD  DSN=PSD.K02.K2D8000.TH8.G4672V00,DISP=SHR,
//   UNIT=(,2)
//DPMPARMS DD  DSN='TTAP.TS2.DB2PM.DPMPARMS',DISP=SHR
//*YSPRINT DD  SYSOUT=*
//SYSPRINT DD DSN=GTS.YV1.DB2T.G4DE.REPORT3,
//**SYSPRINT DD DSN=GTS.YV1.DB2T.G4DE.REPORT1,
//           DISP=(MOD,CATLG,CATLG),
//           SPACE=(CYL,(250,100),,,ROUND),
//          DCB=(RECFM=FBA,BLKSIZE=1330,LRECL=133),
//           UNIT=SYSDA
//*YSOUT   DD  SYSOUT=*
//SYSOUT   DD DSN=GTS.YV1.DB2T.G4DE.REPORT4,
//**YSOUT   DD DSN=GTS.YV1.DB2T.G4DE.REPORT2,
//           DISP=(MOD,CATLG,CATLG),
//           SPACE=(CYL,(250,100),,,ROUND),
//          DCB=(RECFM=FBA,BLKSIZE=1330,LRECL=133),
//           UNIT=SYSDA
//ACRPTDD  DD  SYSOUT=*,DCB=BLKSIZE=133
//SYSIN    DD  *
GLOBAL TIMEZONE(DST)
ACCOUNTING
     TRACE
     FROM      (06/29/01,16:55:00.00)
     TO        (06/29/01,19:05:00.00)
     LAYOUT(LONG   )
     INCLUDE (PLANNAME (QMF610) SUBSYSTEMID(DB2B))
EXEC
//
ACCOUNTING REDUCE
//
//
//
```

90.0   Hangs

"Hangs" and / or potential performance problems are always difficult to investigate. Only through much experience can you understand how to solve these complex problems. First, you should analyze whether there is a bottleneck and if so, where the bottleneck exists.

A bottleneck could exist in any processing layer of the application or DB2. These bottlenecks could include I/O or CPU bound systems, network delay, third party application processing, DB2 processing or others. For each situation, a different set of data needs to be collected. Every problem is unique and uses different layers of connectivity and processing. Because the scope of hang and performance problems requires much experience in debugging, I will only briefly cover some tools and procedures that can be used to help solve these problems.

A hang or performance problem is typically encountered due to bottlenecks at one of the levels described in the following:-

Query Level:- For query level performance problems you can use utilities such as DB2 Explain to generate access plans for specific queries, or you can use DB2 Batch to see SQL prepare and execution timings. You can examine the plans to determine if the optimal access plan is being chosen by the DB2 Optimizer. If not, then specific query tuning will be required to obtain the best results. Because query tuning alone is a very complicated and in-depth topic it is beyond the scope of this paper.

Database and Instance Level Hang:- For these type of performance problems, you can use the DB2 monitoring utilities and stack trace back dumps in conjunction with operating system commands to determine what the problem is. The DB2 monitoring utilities include snap shot and event monitors, and you can use them to analyze a problem over a period of time to narrow down the cause to a specific operation.

If you suspect a hang, then you can use the db2_call_stack and db2ncstack tools. DB2 automatically generates stack "trace back" files when problems are detected that warrant this type of diagnostic information. However, when you want to see the function call stack at a particular point in time for investigating hang situations, the method for doing this varies by platform. On UNIX/AIX DB2, two utilities are supplied to generate stack trace back files:-

1. db2_call_stack generates trace back files for all DB2 processes on all nodes.
2. db2ncstack generates trace back files on a single node.

These commands are issued at an instance level (which means that all databases within the instance may be affected), and all trap files will be created within the DIAGPATH directory.

Use db2_call_stack with caution in a EEE environment, it could take a long time to generate the required files.

91.0   Key Tools

1. db2batch – good for getting the elapsed time of your query and overall database activity.

2. dynexpln a.k.a. Explain – good the for cost of the query(in Timerons) and access path types (includes flow chart).

3. db2advis – The index advisor. Gives you alternative indexes to access the data referenced by the query.

4. db2look – Migrates DDLs from one database to another.

5. db2evmon a.k.a. – The Event Monitor - Monitors Events.

To understand how to use any of the above, in Telnet, key in the name of the function followed by a space and –h.

Example:- If you wish to know how to use db2batch in Telnet, enter at the prompt in lower case:-

db2batch -h

92.0 Performance Tuning Checklist

Nbr.	Item	Description
1	Block Log Full Transactions	Prevent disk "full errors" when log is full – V7.2.
2	Bufferpool	Is the bufferpool the maximum it can be for the tables in the transaction under UDB V7.2?
3	Catalog Tables	Review Catalog tables each table in the transaction / database?
4	Clustered Index	Review Clustering of indexes.
5	Comparisons	When comparing the key columns of two tables, are the keys in the same order? (Datateam only).
6	Configuration Parameters	Review the 'database' and 'database manager' (instance level) parameters.
7	Constraints	Review the constraints on the tables involved in the transaction, follow up with DA.
8	Co-related Names	When the table name is corelated, check columns referenced start with the corelated name (Datateam).
9	Data Distribution	Review data distribution.
10	DB2batch	Apply db2batch analysis results where applicable.
11	Db2look	Apply db2look statistical information where applicable.
12	DDLs	DDL review (Keys vs Indexes).
13	Defaults	Defaults review vs individual settings for tablespaces, bufferpools etc.
14	Erwin	DDL generated from Erwin (DA).
15	Explain	Apply Explain output to SQL (Datateam).
16	Include Index	Use of new INCLUDE feature on Create Unique Index utilized?
17	Log Limits	Review Log Limits increased up form 4GB to 32GB (max). Enables large amounts of work within a single transaction – V7.2.
18	Optimization Class	Review the optimization class adequate.
19	Partitioning Key	Review partitioning key the first column in the relevant tables? (and Partitioning Key Update)
20	Referential Integrity	Review referential integrity based on constraint list..
21	Reorg Index	Review reorgs for tables specify the table name, index name and the tablespace name.
22	Row Blocking	Review use of row blocking in Select statements e.g. OPTIMIZE FOR 1000000 ROWS
23	Runstats	Review reorg/runstats synchronicity
24	Sequence Support	Enables database manager to automatically generate a new numeric value for each call made to the sequence's NEXTVAL – V7.2.
25	SQL	Review SQL complexity. This may show in the Explain results where there are many table scans.
26	Statement Level Isolation	Define isolation levels at the statement level for better granularity and improved performance – V7.2.
27	Static SQL	In V7.2, Static SQL gives better performance than Dynamic SQL if you repeatedly run the same SQL.
28	Table vs Index	Review the sequence of key fields in the table match their definition in the index?
29	Tablespace	Review large tables allocation of DMS tablespaces? SMS is too slow to handle > 1 million rows
30	Timestamps	Review timestamps defined in the index columns of a table.
31	UDB V7.2	Review new V7.2 features Example:- Temporary Tables? Partitioning Key Update?
32	UDB V8.1	Review new V8.1 features. Example:- New Storage Manager, The Memory Tracker, The DB2 Health Center, MDClustering

93.0   Questions and Answers

"What are the meaning of the terms 'sargable' and 'timeron'?"

Stage 1 or Data Manager query processing was originally performed by the RSS (Research Storage System), so named by the System Research group. The DB2 group changed the name soon after the product release. "Sargable" is a contraction of "search arguable." Nobody knows why they, the Research Group, decided to contract and use the latter term to label those predicates that would be handled by the RSS, closest to where the data is stored. Suppose an analogy is "timeron," which is a contraction of "timer on" to specify a unit of query processing/search time, which the Research group also invented for System R.

Where can one find an example of using stored procedures with DB2 and JDBC? :

Access the IBM Redbook "DB2 Java Stored Procedures" (document number SG24-5945). You can find it online at http://www.redbooks.ibm.com/. Once on the Redbooks site, you can enter the document number in the search field to get more information. You can order the hard copy, or access a PDF or an HTML version of the document. Section 3.7 contains client code samples, including an example of a Java client invoking a stored procedure via JDBC.

Q. How can one use different schemas in the same program?

There are quite a few situations where within the same program, different collection IDs or schemas are used to qualify completely different tables.

SQL Stmt #1:

Select Cust From Collid1.tablea
        Where SSN = :HV-SSN.

SQL Stmt #2 in the same program:

Select Store_Num From Collid2.tableb
        Where Store_id = :HV-Store-id

In these situations sometimes the collid has been hardcoded. This can't be good. How does one control which collid to use? I know at run time you can use a packageset to control which schema one connects to. What should one be using for Bind parameters if you don't want to hardcode the collection IDs?

A. Suppose that your production DB2 database is instanced by client ID. That is to say, you have one set of tables for one group of clients, and another set of tables for another group of clients.
The high-level qualifiers used for the two sets of tables are GROUP1 and GROUP2, respectively.

You can set up two collections, also called GROUP1 and GROUP2. All programs use unqualified table names in SQL statements (that is to say, you have SELECT COL1 FROM EMPLOYEE instead of having SELECT COL1 FROM GROUP1.EMPLOYEE). Every program is bound into each of the two collections. When binding program PROGXYZ into collection GROUP1, you specify that GROUP1 is to be used as the high-level qualifier for unqualified table names (this is accomplished by specifying QUALIFIER(GROUP1) on the BIND PACKAGE command).

When binding the same program into the GROUP2 collection, you specify QUALIFIER(GROUP2) on the BIND PACKAGE command.

At execution time, you use SET CURRENT PACKAGESET to determine the table to be accessed. Suppose, for example, that you need to access data for an employee of a client in the GROUP1 set of clients. In your program, you'd have a SET CURRENT PACKAGESET = :HVAR, and you'd place GROUP1 in the host variable. Then, you'd have your select-statement, for example, SELECT NAME, ADDRESS FROM EMPLOYEE WHERE... Because you have pointed DB2 to the GROUP1 collection, DB2 will look in that collection for the package corresponding to the program. It will find the package because all programs are bound into all collections. Because QUALIFIER(GROUP1) was specified for that package at the time it was bound into the GROUP1 collection, the above SQL statement will actually be executed as SELECT NAME, ADDRESS FROM GROUP1.EMPLOYEE WHERE... Thus, picking the right table is as simple as picking the right collection.

Q. How can I create new database objects similar to existing objects in DB2 UDB?

A. You can create new database objects similar to existing objects in DB2 UDB by using a tool (program) called DB2LOOK. With DB2LOOK you can extract DDL from an existing DB2 database. You can not only extract the DDL for a given database, but also it's statistics by telling the DB2LOOK program to build the necessary SQL UPDATE statements to update the statistics in the DB2 Catalog. This is a terrific way to clone a database for testing purposes.

Q. In DB2, how can I create my own specialized datatypes?

A. You can create datatypes, called distinct types, to enhance the use of DB2 built-in datatypes. This feature has been available for some time in UDB and became available in OS/390 since Version 6. A distinct type will be based on a DB2-supplied datatype (source type) and it will have uniquely defined rules for comparisons and calculations that are applicable to your data. For example, to create a distinct type called CURRENCY based on the decimal datatype, you could use the following statement:-

CREATE DISTINCT TYPE CURRENCY AS DECIMAL(6, 2) WITH COMPARISONS

The CURRENCY is based on the decimal datatype, but it will only be comparable with another occurrence of the same distinct type. By using the WITH COMPARISONS clause, you are telling DB2 that this distinct type can be used in comparison operations with other occurrences of data using the same distinct type. The names of the distinct types must be unique within a schema.

Distinct types are kept in the DB2 catalog called DATATYPES and can be given a description in this table. These datatypes can now be used by a CREATE TABLE statement or an ALTER TABLE statement.

Q. If I reorganize a tablespace, does it automatically reorganize the corresponding indexes?

A. Yes. If you reorganize the entire tablespace, the indexes will be rebuilt during the BUILD phase of the REORG utility. If you reorganize only a partition of a tablespace, all of the nonpartitioning indexes are affected, but they are not rebuilt. They are simply corrected in the BULID2 phase and these nonpartitioning indexes may require a REORG after the tablespace partition-level REORG.

Q. How much temporary disk space is required to reorganize a table in the UDB environment?

A. The amount of temporary space required by REORG is about the same required by the original table.

Q. We are constantly having Lock Escalation on our Database. Our Lock parameters are Locklists 1600 and MAXLOCKS 16. Please explain the DB2 formula for dealing with locks and please list the best parameters we can use to minimize lock escalation.

A. The LOCKLIST parameter identifies the number of pages available in the LOCKLIST HEAP. MAXLOCKS identifies the percentage of the total of all locks used by an application before LOCK ESCALATION "kicks in". With your parameters, 1600 page locklist and Maxlocks set to 16, escalation will be triggered when any one application holds over 700K of locks. Each lock consumes up to 72 bytes (the first lock takes up 72 bytes and the second and subsequent locks take up 36 bytes). Consequently you are allowed 8630 – 72 byte locks and 17256 – 36 byte locks per application connection.

Because you are hitting lock escalation frequently, you may want to change/modify the applications to COMMIT more often. Also, your application may need a redesign as you are well over the average number of locks per application as recommended by IBM, which is 512!!!

Here are some parameters in common use which may help reduce your Lock Escalation poroblems.

LOCKLIST	= 4096 (Max Storage for lock list in 4KB Units)
DLCHTIME	= 5000 (Deadlock checking interval)
MAXLOCKS	= 90 (Percentage of lock lists per application)
LOCKTIMEOUT	= 180 (Lock time out in seconds)

CHNGPGS_THRESH = 40 (changed pages threshold)

To reduce lock escalation in general, increase the commit frequency, reduce RR Isolation Level, and/or add LOCK TABLE Statements to the Applications

To find out the percentage of LOCKLIST used in your particular databases use the following script:-

```
LOCKLIST='db2 get cfg for $db|grep LOCKLIST | awk'{Bytes = (($9*4)*1024)}
((printf("%d",Bytes)} {exit}"
LOCKLUSE='db2 GET SNAPSHOT FOR DATABASE ON $db \
|grep "Lock list memory in use (Bytes)" | awk '{print $8}"
```

If the percentage LockList in use is greater than 60% of the MaxStorage size, then performance will degrade because of lock escalation, on the other hand, if LockList
is too low it may be over allocated.

94.0   REORGS Revisited

Command Syntax
```
>>-REORG TABLE--table-name----+-------------------+----------->
                              '-INDEX--index-name--'


>-----+---------------------+------------------------------><
      '-USE--tablespace-name--'
```

Command Parameters
TABLE table-name
Specifies the table to reorganize. The table can be in a local or a remote database. The fully qualified name or alias in the form: schema.table-name must be used. The schema is the user name under which the table was created.
INDEX index-name
Specifies the index to use when reorganizing the table. The fully qualified name in the form: schema.index-name must be used. The schema is the user name under which the index was created. The database manager uses the index to physically reorder the records in the table it is reorganizing. If the name of an index is not provided, the records are reorganized without regard to order.

If the name of an index is specified, the database manager reorganizes the data according to the order in the index. To maximize performance, specify an index that is often used in SQL queries. If the name of an index is not specified, and if a clustering index exists, the data will be ordered according to the clustering index.

CLUSTERRATIO or normalized CLUSTERFACTOR (F4) will indicate REORG is necessary for indexes that are not in the same sequence as the base table. When multiple indexes are defined on a table, one or more indexes may be flagged as needing REORG. Specify the most important index for REORG sequencing.

From the Staging Box (10.3.66.34) (Prod 008:-

cd /stg/dba/e8412psg/trigger reorg staging

cd /prd/dba/e8412pol/trigger reorg production

95.0 Rows per Node in a Partitioned Table Space

```
-------------------------- Command entered --------------------------
SELECT NODENUMBER(METCNCT_CUST_ID), COUNT(*) FROM TOLCUST_
CNTRC_NM GROUP BY NODENUMBER(METCNCT_CUST_ID)

-----------------------------------------------------------------------

1          2
----------- -----------
          3    6470908
          5    6405390
          7    6417207
          9    6392584
         11    6402600

  5 record(s) selected.
```

Where:- METCNCT_CUST_ID is the partitioning key.

96.0   Runstats

Runstats are important, and it is important to maintain statistics in the catalog. Not only does the Optimizer use these figures, but the DBA should use these figures to alter objects or run utilities when certain thresholds have been reached.

In V7 there are more Runstats statistics gathered. An important one is the number of extents on the disk.

For a DBA, not only are the current figures important, but also a history of these statistics is important. Using history, you can see growth and also pinpoint changes in statistics relating to the changes in the environment.

Third party vendors not only created their own version of Runstats but also provided the DBA with a history. In V7, the Runstats utility can also collect a history. This action cannot be enforced, and the default is NOT to collect a history, so you have to change your Runstat statements in order to use this enhancement.

The REORG and REBUILD utilities can also run Runstats in parallel and have also been changed to allow for the new History keyword.

Example:- REORG INDEX ....HISTORY

RUNSTATS TABLESPACE .... HSITORY

If you collect RUNSTATS with the new HISTORY keyword, then the new statistics are also written in the _HIST tables. There is also an option to choose NOT to update the normal tables used by the optimizer (UPDATE NONE) but still write a history (HISTORY ALL). Using this great feature you can query the history tables and create your own "trend analysis reports using some type of Report Writer.

97.0   The Historical Statistics

The Historical Statistics are kept in nine different DB2 Catalog Tables.

- .SYSCOLDIST_HIST
- .SYSCOLUMNS_HIST
- .SYSINDEXPART_HIST
- .SYSINDEXES_HIST
- .SYSINDEXSTATS_HIST
- .SYSLOBSTATS
- .SYSTABLEPART_HIST
- .SYSTABLES_HIST
- .SYSTABSTATS_HIST

98.0   Savepoints.

SQL statements can now be grouped into an executable block. If any SQL statement in the block (within the SAVEPOINT scope) returns an unsatisfactory SQL code, then the application may elect to rollback to SAVEPOINT. In the unlikely event of a statement or database manager crash, the entire unit of work will still automatically roll back to its start. Once all statements in the block have completed, execution can continue or can be rolled back to the start of the SAVEPOINT. The new SQL commands SAVEPOINT and RELEASE SAVEPOINT carry out this new function, which gives application developers greater granularity of control over units of work.

99.0   Stored Procedures

Stored procedures. Version 7.1 includes a number of new features for stored procedures that facilitate the application development process, including:

- **Nesting**. A stored procedure can now invoke another stored procedure, facilitating reuse, manageability, and code modularity.
- **Building**. Enhancements to the Stored Procedure Builder tool, originally introduced in version 6.1 (generally considered a flop) , simplify stored procedure management, development, and testing. Simply click Start -> Programs -> IBM DB2 -> Stored Procedure Builder to create stored procedures in a matter of minutes. A SQL Wizard even makes building SQL statements simple.
- **SQL in stored procedures and user defined functions (UDFs).** The SQL99 standard provides SQL language statements that can be used in stored procedures. SQL3 defines control statements for use in SQL functions. Version 7.1 and above lets you build stored procedures that fully exploit and adhere to these standards:
- BEGIN ... END
- Local variable declarations
- Value assignments
- CASE statements
- IF statements
- Loops (LEAVE, FOR, WHILE, REPEAT)
- RETURN.

The new SQL Procedure Language provides a very simple and powerful alternative language for creating stored procedures.

Anyone who ever wished for the ability to include SQL in user-defined functions might want to try version 7.1. Scalar, row, and table functions can now contain SQL.

100.0 Migrating

Migrating to DB2 UDB version 7.1. In addition to the enhancements for migrating from other DBMSs to DB2, version 7.1 makes it easier to migrate from older DB2 versions to the new release. Anyone who uses DB2 UDB for Unix, Windows, and OS/2 will be pleased to learn that version 7.1 and above supports version skipping. Generally, DB2 UDB version 5 and version 6.1 users can use migrate with provided facilities directly to version 7.1. This extraordinary, version-skipping support for jumps from DB2 UDB versions 5 and 6.1 to version 7.1 enables customers to span a three-generation gap in DB2 UDB.

John V. McLean

101.0   Relation Table Scans

As the size of the table grows, the length of the actual time taken to scan the entire table grows exponentially once the size of the table exceeds the amount of memory available to contain the pages from a relation scan. The significant increase in time is due to page faulting in memory.

Sometimes the immediate availability of data is an absolute must. However, there are other times when there is a need to load data into a table. In the past, ways of getting data into a table were load, import or write a user program to perform inserts. This makes the data unavailable. Writing a user program also takes time coding binding etc, not a user friendly solution. There are however some advantages using SQL INSERTs instead of the LOAD utility. For example, INSERT Triggers will fire, something LOAD does not do. DB2 also optimizes the insert process by creating and using an I-PROC) a kind of small program that actually does the insert) for reuse by the next INSERT.

The solution to these problems arrives with V7 and above in the form of the new SHRLEVEL CHANGE. With this new SHRLEVEL the LOAD will do normal INSERTs. It does CLAIM processing and no DRAIN (see below) as a normal utility does. All related activities occur as they would with SQL:- RI, Triggers, Duplicate Keys, Free Space processing etc. This is YGWYAF (You Get What You Ask For) processing. The BIG bonus:- 100% availability of data and easy to use (no need to code a program). The only things that are different from SQL processing are that you need LOAD privilege (not INSERT privilege) and that the utility uses a utility timeout instead of a SQL timeout. Of course the LOAD does an optimized load by interfacing directly to the Data Manager so no SQL PREPARE or authorization check is needed. Commit processing is also handled by the utility. Also, the LOAD will constantly monitor the locking situation and adjust the commits based on this scenario. Restarting is no problem since the utility takes checkpoints.

Once we are doing SQL INSERT processing, what is stopping you from doing an INSERT into a table on another server using DRDA? Nothing! IBM has mastered this feature and the end result is a cross-loader technique that allows you to load data over a network.

CLAIM:- DB2 uses a *claim* to register that a resource is being used. Claims are basically usage indicators. A process stakes claim to a resource, telling DB2 that it is using the resource.

DRAINS:- DB2 makes sure that all claims to a resource are "gone" before a new claim is established for the resource. The mechanism of removing old claims and preparing for new ones is done by *drains*. A drain places locks on a resource, before it can remove the old claim off the resource and release the resource, making it ready to be re-acquired by a new claim.

102.0   Problem

RS6000 server with multiple CPUs that are running 100% CPU all day. There are > 7 DB2SYSC processes active which are consuming all the CPUs. NO memory shortage. I/O is low. User is running EEE 7.2 on AIX 4.3.3. Response time is dreadful. No db2diag.log!!!

Solution:-
You could be doing table(space) scans in your bufferpools, I see this very frequently when we do performance audits and reviews.

<u>Steps I suggest:</u>

1) db2 get snapshot for database on DBNAME
    a) Add the number of commits plus rollbacks to get total transactions

2) db2 get snapshot for tables on DBNAME
    a) For each table, Divide the number of "Rows Read" by Total Transactions to find Rows Read per Transaction.
    b) As a rule of thumb, if RR/TX > 10, could be a SQL or index tuning opportunity.
    c) If RR/TX > 100, big tuning opportunity
    d) If RR/TX > 1000, severe degradation exists

3) Find the SQL that is causing the high RR/TX. Several methods are available. In my opinion, the best is DGI's SQL-GUY(TM) or Flight Deck for DB2 - these tools use DB2 SQL Event Monitors to accurately capture, score, and cost rank all of the SQL through the DB2 engine _without_ having to do any I/O to tables. In lieu of a vendor tool, capture SQL event output to a file then process it with 'db2evmon', then scan the voluminous output for the table names having high RR/TX – this latter technique only works if the table names are in the SQL.

103.0 Mainframe Tip

If you need to know how many VSAM EXTENTS have already
been allocated, use this SQL. Remember to run STOSPACE and
RUNSTATS first:

```
-------------------------------------------------------------
----------
--                 T A B L E S P A C E S
-- THIS QUERY MONITORS THE SPACE ALLOCATED AND THE SPACE USED FOR
-- FOR EACH TABLESPACE. THE "STOSPACE UTILITY" SHOULD BE RUN BEFORE.
-------------------------------------------------------------
SELECT T.DBNAME, T1.NAME, T.SPACE, T.PQTY*4 AS "PQTY",
       T.SQTY*4 AS "SQTY",
       (T.SPACE - (T.PQTY*4)) / (T.SQTY*4) AS "# EXT",
       (T.SPACE*100)/(T.PQTY*4) AS "% PQTY USED"
       FROM SYSIBM.SYSTABLEPART T, SYSIBM.SYSTABLESPACE T1
       WHERE T.DBNAME    = 'DSNDB06'
       AND  T1.DBNAME    = T.DBNAME
       AND  T1.CREATOR   = 'SYSIBM'
       AND  T.TSNAME     = T1.NAME
       AND  T.STORTYPE = 'E'
       AND  T.PQTY      > 0
       AND  T.SPACE     > 0
       ORDER BY T.DBNAME, T1.NAME;
-------------------------------------------------------------
 SAMPLE RESULTS:


---------+---------+---------+---------+---------+---------+-------
NAME        SPACE       PQTY       SQTY     # EXT  % PQTY USED
---------+---------+---------+---------+---------+---------+-------
XXX0001      7200       7200       1440       0       100
XXX0002      3600       3600        720       0       100
XXX0004     72000      72000       7200       0       100
XXX0005    864000     720000      72000       2       120
XXX0007     72000      72000       7200       0       100
XXX0008    720000     720000      72000       0       100
XXX0009      1440       1440        720       0       100
```

104.0   Special Registers

Special registers contain values maintained by the system, that describe and control the environment in which an SQL statement is executed.

Listed are some of the most commonly used special registers.

Register Name	Data Type	Description
CURENNT DATE	Char(5)	Contains the date on which the current statement is being executed.
CURRENT DEGREE	Char(5)	Used to set limits on the degree of intra-partition parallelism to be used by dynamic SQL statements.
CURRENT EXPLAIN MODE	Char(8)	This register controls the gathering of Tabular Explain data for dynamic SQL statements. Tabular Explain describes the access plans chosen by the optimizer For a given SQL statement.
CURRRENT EXPLAIN SNAPSHOT	Char(8)	This register controls the gathering of Snapshot data for dynamic SQL statements.
CURENT NODE	Integer	Used as an identifier when used in a parallel database system with multiple partitions. It's node number is equal to the node number to which your application is connected.
CURRENT QUERY OPTIMIZATION	Integer	This register specifies the class of optimization techniques to be used in preparing Dynamic SQL statements.
CURRENT SERVER	Varchar(18)	Contains the name and address of the database to which your application is currently connected.
CURENT TIME	Time	This register contains the time of day at which the current SQL statement is being executed.
CURRENT TIMESTAMP	Timestamp	Contains a microsecond measure of the date and time at which the execution of the Current SQL statement began.
USER	Char(8)	This register contains the User Id. Of the user who is connected to the database And executing the current application.

105.0   DB2 UDB Tools

db2batch – Gives performance information for execution timings and result set analysis.

db2bfd – Provides a detailed description of the contents of a bind file.

db2cat – Analyzes the contents of a tables packed descriptor.

db2ckbkp – Checks a single DB2 backup image or multiple parts of an image.

db2dart – Verifies that the architectural integrity of a database is correct.

db2drdat – Traces the DRDA data flows exchanges between servers and requestors.

db2flsn – Returns the name of the relevant log file.

db2ipxad – Returns the DB2 server's IPX/SPX internet work address.

db2level – Displays detailed output about the level of DB2 including Fix Pack Level.

db2look – Extracts the DDL(s) necessary to re-create database objects.

db2move – Performs mass data movement from one database to another.

db2recri – Re-creates indexes that were marked invalid during restart.

db2tbst – Provides a text description of a given table space.

db2trc – Traces all functions within the database manager per node in EEE.

db2untag – Removes the DB2 tag from a table space container.

106.0   DB2DART

db2dart verifies that the architectural integrity of a database is correct and has the ability to repair some database damage.

This tool confirms the following:-

1. That the control information is correct
2. That there are no discrepancies in the format of the data
3. That the data pages are the correct size and contain the correct column types.
4. That the indexes are valid.

db2dart can be run at the table, table space or database (default) level, and can also dump data and index pages. If you use db2dart with the /DDEL option, it becomes a very powerful data recovery tool. It can dump all data for a table into delimited ASCII files.

107.0 Joins and Star Joins

Joins

Joins "join" tables and/or views together to enable SQL Queries to get their answers in "one shot". Thus a join can serve you well in certain circumstances, on the other hand too many joins can cripple your system and ensure that your query results never materialize!!

IBM recommend no more than six or seven tables in a join. But I have seen users want to join thirty or forty tables in a join, the result being they wait forever for the data to materialize on the screen. Sometimes the query runs all night long with no materialization comes morning. This of course is a complete waste of time. It is better to reduce the number of tables or views in a join to get answers rapidly in a series of joins, rather than wait forever by doing a join with many tables. This is a trial and error process, and can be helped by using a measuring tool such as Load Runner, which will pinpoint problems with joins, giving start and end times and a host of other useful information which you can use to guide you make workable joins.

Here is how it works:-

Tables that are joined usually have something in common, like parts of a composite key, but each table in a join obviously does not have all the information needed for the query, hence the need to join.

Debugging a Join

Sometimes, you do not get results even when joining a few tables, and you seem unable to fathom why!! In these cases, comment out all the SQL statements in the join except the last one and run, if still no good then scrutinize the last SQL equation. However if you get "success" then uncomment out the next but last SQL equation and run, scrutinize the results. Keep going like this (un-commenting the SQL statements) until you do not get a "hit". Then, this is the SQL equation that is causing the problem investigate and solve!!

Star Joins

Since the introduction of the Star Join access method in Version 6 of DB2 UDB, there have been many additional enhancements to improve the detection of star schema queries to determine the most efficient access path, and also to minimize regression of non-star schema queries.

Due to the rapid deployment of these improvements, the challenge has been to ensure that the DB2 community is aware of this enhanced functionality. Thus, it is the objective of this page to provide an overview of the issues and complexity associated with this join methodology.

Star Schema

A star schema is the typical data model used for business intelligence and data warehousing systems, where there exists a number of normalized tables (known as dimensions) which surround a large centralized table known as the Fact Table. The Fact Table stores the fact, such as sales transactions, and the dimensions store the attributes about these facts. The dimensional tables are usually depicted as surrounding the fact table, hence the term star, each dimensional table representing the point of a star (does not have to be a three or five pointed star!)

Star Schema join challenges

In an OLTP world, the query optimizer technology has been designed to handle pair wise joins; joining two tables at a time, based on join predicates and determining the join sequence based on associated cost. A join pair is formed by the existence of join predicates between tables, thus tables without join predicates are considered to be unrelated and therefore not valid join pairs.

The pair-wise join method fails miserably for star schemas. As dictated by the star schema model, the only table directly related to other tables is the Fact Table. As a consequence, the Fact Table is most likely chosen in the first pair-wise join, with subsequent dimension tables joined based on cost.

Even though the intersection of all dimensions with the Fact Table can produce a small result set, the predicates applied to one single dimension table are typically insufficient to reduce the enormous the number of Fact Table rows.

If a join based on related join-pairs does not provide adequate performance, then an alternative is to join unrelated tables. Joining of unrelated join-pairs results in a Cartesian product, whereby every row of the first table is joined with every row of the second. Performing a Cartesian join of all dimension tables before accessing the Fact Table may not be efficient. The optimizer must decide how many dimension tables should be accessed first to provide the greatest number of filtering of Fact Table rows using available indexes. This can be a delicate balance as further Cartesian products will produce a massive increase in the size of the intermediate result sets. Alternatively, minimal pre-joining of unrelated dimension tables may not provide adequate filtering for the join to the Fact Table.

108.0   Stored Procedures revisited

Stored Procedures (SPs) store business logic inside the database. An SP is a callable program stored in DB2 at a local or remote server which can execute SQL statements. SPs reside in their own address space or if Workload Managed in multiple address spaces and can be executed by many callers in parallel. SPs can be written in many languages including REXX, COBOL, C, Assembler, SQL and Java. An SP will normally have at least one SQL statement.

SPs offer a range of benefits for e-business applications such as:-

Reduced network traffic, translates into increased throughput – An SP can issue many SQL statements, so the number of network send/receive operations are minimized, this improves the elapsed time and CPU time of the application. The SQL issued by the SP uses CALL Attach or RRS Attach, so there is no added distributed overhead on the SQL statements. For applications that issue many SQL statements the savings on Elapsed and CPU time can be quite significant. Application time is reduced as the SP runs on the database server.

Static SQL (SQLJ) can be used in Java Stored Procedures – This translates into better performing SPs. With static SQL (SQLJ), the syntax is much simpler than JDBC. For instance, to do a simple select using JDBC you require four statements (1. Prepare(), 2. Execute Query(), 3. Next(), and 4. Close(). When using SQLJ you would code a single SQL statement (select * from mycreator.mytable).

Business Logic in Database Server – SPs run directly on the server, they can take advantage of the extra memory, processing power and they are much closer to the data so network traffic is kept to a minimum. Change management is another selling point for SPs, because they reside in one central location, should any changes need to be made; they are made in one place effective immediately.

Security – Owing to the fact that SPs run in a secure DB2 address space it is extremely difficult for outsiders to update or change any logic in the SP. SPs also use static SQL to access DB2 tables. The SP uses the privileges of the person who bound the SP to DB2. This in itself removes the need to grant access of tables to end users, which makes managing access easier because the client applications do not need explicit access to DB2 tables.

Easier to control locking contention – If the SQL against DB2 is being executed from DB2 then you can control the locking and expedite the work being done because you do not have to wait application think time or network response time. If for example the client application issues an SQL select against DB2 on a mainframe, a thread is opened, a lock is taken, data is sent down to the client and then DB2 waits while the client performs some logic, then the client application requests an update which DB2 performs and the application performs some more logic and terminates. This is seen as wasted time because if a SP, UDF or Trigger had executed the SQL statement, the unit of work would have been executed fasted, because most of the time spent waiting for the network to deliver the data and messages to the program and for the program to act on these would be eliminated. Therefore because the task being done finishes a lot quicker, the threads and locks are released a lot sooner, so you can get more done is less time or do more with the same resources.

Integration with desk top tools – SPs integrate with desk top tools and are supported by ODBC, JDBC and SQLJ.

Communication with Stores Procedures – There are many ways of passing data to and from a stored procedure. The trick is to use the most appropriate mechanism for what you are trying to achieve. Communication with SPs can take three forms:-

1. Simple parameter passing – where single or multiple parameters can be passed. Use locators to fetch from stored procedure cursor – SPs can return the result set and the parameters specified in the CALL statement to the client. This allows the client application to fetch the rows in the result set, almost as if the client itself had the issued the SQL statement. A result set could be imagined as a table of 0 or 'n' rows. The process to retrieve these rows is in two steps a). The calling program calls the SP, which opens a result set cursor and returns to the calling program (with the results set still open); b) The calling program allocates a cursor to the locator and then the calling program fetches the allocated cursor, returning rows from the results set, a row at a time.

109.0 E-Commerce

The e-commerce business environment is harsher than the Australian outback. A clear understanding of customer behavior can help you and your company come out on top.

E-commerce gives companies the opportunity to increase revenues by opening access to worldwide markets and improving profitability. It also opens the door to a host of customer loyalty challenges. As the competition to win Web customers becomes fiercer, insight into customer characteristics, purchasing and navigation behavior becomes even more valuable. Understanding customer characteristics and behavior helps companies design and implement more effective, cost efficient Web sites, marketing campaigns, and product promotions.

The enormous amount of 'click stream' and transaction data generated on e-commerce sites is key to achieving this kind of understanding. The data provides opportunities for business to improve their product planning, marketing strategies and customer satisfaction. But finding answers in raw data requires sophisticated analysis. Many vendors are developing out-of–the-box products for analysis to meet this need. Such technologies the best of which provide:-

State of the Art data mining and OLAP

Near real time reports that business users can understand

High performance scalable database techniques that support analysis optimization

Data warehousing technologies that facilitate consolidation and cleansing of multi channel data.

IBM provides a subset of this important functionality in it's entry level Web Sphere Commerce Analyzer (WCA), which is included with the Web Sphere Commerce Suite (WCS) V5.1 Pro Edition, designed to analyze and report statistics generated by your customers.

WCA automatically extracts data from WCS to a separate dedicated machine and processes the database records to compile pre-built about traffic and usage. These reports answer such questions as:-

How many times did a shopper click on an initiative and then order the product suggested?

What is the sales revenue breakdown by geography, demographics and time?

How many shopping carts have been abandoned?

What patterns and relationships exist is data that are not obvious?

110.0   Unique Constraints

There are Unique Constraints – rules used by the Database Manager to enforce uniqueness in the values of a key.

There are Referential Constraints – such as *Foreign Keys* whose values are required to match at least one primary key or unique key value of a row in it's parent table.

There are also Table Check Constraints – this constraint is a rule that specifies the values allowed in one or more columns of every row of a table. Table check constraints are defined using the CREATE TABLE and ALTER TABLE statements. For example,

ALTER TABLE <table name>
ADD CONSTRAINT <constraint name>
CHECK (<check condition>)

The third line of the example above is where the check constraint is defined. A check condition might be that an employee's base salary may not be less than a certain amount (CHECK(basepay>30000)) or that the hire date is not before a certain year.

A table can have any number of table check constraints. The constraints are enforced when a row is inserted into a table or when a row is updated. A check constraint defined on a table automatically applies to all sub-tables of that table.

A table check constraint is a restricted form of a search condition. A search condition specifies a condition that is true, false or unknown about a given row. The result of a search condition is derived by applying the specified logical operators (AND, OR, NOT)
to the result of each specified predicate; if logical operators are not specified, the result of the search condition is the result of the specified predicate.

Similarly, a table check constraint is enforced by applying it's search condition to each row that is inserted or updated. An error occurs if the result of the search condition is false for any row.

When one or more table check constraints are defined in the ALTER TABLE statement for a table with existing data, the existing data is checked against the new condition before the ALTER TABLE statement succeeds. Alternatively, the table can be placed in a check pending state by using the SET INTEGRITY statement, which allows the ALTER TABLE statement to succeed without checking the value of the data. The SET INTEGRITY statement is also used to resume the checking of each row against the constraint, it works on a "flip flop" switch principle.

A check constraint specified as part of a column definition applies only to that column. A check constraint specified as part of a table definition can have column references identifying columns previously defined in the same CREATE TABLE statement. Check constraints are not checked for inconsistencies, duplicate conditions, or equivalent conditions. Consequently, contradictory or redundant check constraints can be defined, resulting in possible errors at run time.

111.0 Tablespace Examples for future use

111.0.1   Temporary Tablespace

The TEMPSPACE1 tablespace is defined in the IBMTEMPGROUP Node Group.

If you do not specify any tablespace parameters with the Create Database command, the database manager will create these tablespaces using SMS Directory Containers. These Directory Containers will be created in the subdirectory created for the database, the extent size for these tablespaces will be set to the default.

The naming schema for Unix is:-

Specified_path/$DBINSTANCE/NODEnnnn/SQL00001

To create a temporary tablespace:- CREATE TEMPORARY TABLESPACE TEMPSPACE2 MANAGED BY SYSTEM USING ('d')

Then drop the old temporary tablespace supplied by the system:- DROP TABLESPACE TEMPSPACE1

Also you may try using DMS and specifying your own space parameters.

112.0 With IBM z-Series and z/OS: IBM extends leadership.

What if tomorrow, fifty times more people need to access your Web site than they do today. Could your servers cope? Are you ready for the next generation of e-business?

With less than 2% of the world on the Web and with the explosive growth of pervasive devices, then fifty times may be a conservative estimate. In this new world data and transactions explode. To manage this you are going to need different types of servers. They will need to do more and do it more intelligently. It's an exciting world we live in, and tomorrow will be even more exciting, but excitement brings challenges, daunting ones.

You need new tools for managing e-business
You need total application flexibility
You need innovative technologies

And where do you need these most? In the Information Technology Infrastructure that supports everything you do. IBM's z-Series, powered by z/OS, is the first e-business enterprise server designed for the high performance data transaction needs of the next generation e-business. And the z800 and z900 just got faster and more flexible, offering expanded internal networking capabilities and new virtualization possibilities.

z-Series servers can deliver the highest level of application availability required in today's global networked environment. Even in a single footprint, z-Series servers are designed to avoid or recover from failures to minimize business disruptions. High availability is realized through very high component reliability and design features that assist in providing fault avoidance and tolerance, as well as permitting concurrent maintenance and repair.

For even higher levels of availability, the superior choice is z-Series 900 with IBM Parallel Sysplex clustering technology. New faster Coupling Links provide balanced performance for the powerful z900 in a sysplex. ISC -3 links environment provide up to two gigabits/ second transfer rates; ICB links provide up to one gigabyte/second. In addition, complete backward compatibility exists with S/390 ISC and ICB links.

Another aspect of availability is non-disruptive growth, enabled in the z-Series by the Capacity Upgrade on Demand facility.

113.0 DB2 Tools

Ten thousand tables, thirty thousand indexes, dozens of terabyte databases, sounds like something out of Kafka? No, it's the reality of today's world! How does one manage these objects, tune applications and subsystems and keep up with the latest information on Java, Linux and other "themes"? How does a company survive if these objects are in less than A1 condition? DB2 objects are the foundation of the application, be it a stand-alone zSeries application or a multi-tiered Wed-based ERP application? If these objects are rotten, like a house foundation, the house will come tumbling down.

The information about the care and feeding of these objects is based on knowledge in the DB2 Catalog. Yet the catalog itself has evolved over time and continues to gain in complexity just at a point when we find ourselves with a strong desire to simplify (remember KISS?). With additional tables and columns and columns that replace columns, how can any one person keep track of exactly where the pertinent information is within the catalog?

Finally we find ourselves managing multiple environments, an environment for each development team, unit test, stress test, warehouse test etc. In addition, some DBAs need to manage objects in distributed systems. The logistics of handling these environments is mind-boggling. Mistakes get made, and in the worst case production systems wait for us to fix things.

What if there was a way to automate the understanding of the DB2 Catalog, reduce the amount of time spent managing DB2 objects, improve the accuracy of managing these objects and determine problems ahead of time? In addition what if you could use this information to interactively rectify these problems?

The answer to these questions is found in DB2 tools – specifically administration and database analysis tools. There are numerous DB2 tools now available, and I refer you to the IBM manuals or Web site to get a list of these. The first benefit of using tools as opposed to doing things manually is getting information from the DB2 catalog. With Version 7 there are now 81 catalog tables of which 21 are new. There is an average of 14 columns per table. Out of these 81 tables, 21 of them can have rows inserted, updated or deleted. Now where do you find the find the information that you need to address the task at hand? How many times have you written an SQL statement to get information, only to realize that you've forgotten that the index name is IXNAME in one table and INAME in another? How many times have you needed to write complex or repetitive queries to get the information you're after?

There are tools that will help you select the object information you require and then present the information to you in readable format. Simple commands can help you
drill down to other interesting points about tables. Simple commands can help you find the columns or indexes to a particular table. From these types of drill-down and informative menus, you can move to other options that are important in your quest for catalog information.

114.0 Triggers

A trigger comes into effect when you insert, delete or update data in a table. A trigger is an action made up of one or more SQL statements executed whenever a "triggering event" occurs on a specified table. A triggering event is always caused by an SQL statement (insert, update or delete) which can be executed dynamically, statically, locally or remotely. The triggering event is monitored by DB2, and whenever it occurs the trigger is executed or "fired".

Triggers can only contain SQL statements, not program logic unfortunately, but that does not restrict their capability at all. You can still execute Stored Procedures and User Defined Functions, which means you can "kick off" thousands of lines of code from a single trigger.

Triggers are very useful as they can be used to perform many automatic actions, such as auditing data changes, cascading changes through related tables, enforcing column restrictions, maintaining the referential integrity across a database, enforcing complex
business rules, automatically generating derived column values, and even enforcing e-mail alerts.

There are two types of triggers: After Triggers and Before Triggers. After Triggers perform actions after the SQL statement has completed, and Before Triggers execute before an insert/ update/delete commences. A Before Trigger can decide whether to allow or reject insert, update or delete statements. For example, someone buys a book and before the transaction is processed, the Before Trigger validates that the bookstore's international couriers deliver to the country of the client. You can use all SQL logic in Before Triggers to determine if the action is allowed. So you can check dependent tables as well as constants. Triggers amongst other things are good for removing the need for business validation at the client end. A few things a Before Trigger CANNOT do are, cause further updates to the database either directly or indirectly and activate another trigger.

Triggers can perform actions at the statement level (FOR EACH STATEMENT) or at the row level (FOR EACH ROW). Triggers that are specified with FOR EACH STATEMENT are triggered only once for the triggering SQL operation. Triggers that are specified with FOR EACH FROW execute for each row of the table that the triggering SQL changes. If the triggering SQL does not change any rows, the trigger is not executed. You can use triggers to track the before and after values of sensitive data changes (e.g. salary information) and specify CURRENT USER or CURRENT TIMESTAMP special registers in the SQL statement to capture this information, (because triggers work at the table level they are completely transparent, so people will not be aware that these triggers even exist).

Trigger Example:- CREATE TRIGGER SALARY_CHK AFTER UPDATE OF SAL_TABLE ON EMP_COL REFERENCING OLD AS OLD_EMP_REC NEW AS NEW_EMP_REC FOR EACH ROW MODE DB2SQL WHEN (NEW_EMP_REC.SALARY > (OLD_EMP_ REC.SALARY*1.3)) BEGIN ATOMIC SIGNAL SQLSTATE '75001' (INVALID SALARY CHANGE);

115.0 DB2 UDB V8.1 ESE General

The beta code is in English only and supports both 32 and 64 bit versions for AIX, HP, Linux, Solaris and Windows. The beta release of IBM's DB2 V8 for Linux, Unix and Windows marks the next stage in the evolution of relational database providing for new levels of information integration within your e-infrastructure. DB2 V8.1 allows you to access any information, from any application, from any place in your organization anytime, and still at the total lowest cost of ownership in the industry.

> **DB2 V8.1 builds on the strengths that customers and partners have come to expect from IBM data management. This newest release provides new and significant SMART (self-managing and resource tuning) automation capabilities, Business Intelligence enhancements, and expanded abilities to integrate information across your company wherever it resides. DB2 V8 also builds on the capabilities of reliability, performance and scalability you have come to expect from IBM. Highlights of the new capabilities in DB2 V8.1 are outlined below.**

Self-Managing Capabilities

DB2 V8 provides significant automation capabilities including self-configuring, self-healing, self-optimizing and self-protecting. Through new tools and wizards DB2 reduces errors and increases performance. DB2 V8.1 eliminates, simplifies, monitors and automates many database functions, notifying the DBA proactively, in the event a problem is about to occur and recommend corrective actions to resolve the issue.

Business Intelligence

To help you gain a faster insights and ROI from your data DB2 V8.1 incorporates sophisticated Business Intelligence capabilities that allow you to easily reorganize stored information to perform faster using more insightful queries. The new feature called multi-dimensional clustering enables customers to reorganize information in one easy step and perform simultaneous searches on it. By clustering data DB2 can perform queries and analytics nearly 100% faster than before.

Information Integration

DB2 V8.1 helps you solve critical business problems by integrating information across your company and minimizing the tasks and costs associated with maintaining your database infrastructure. DB2 has unique federated data management capabilities that allow you to access manage and analyze your data wherever it resides. DB2 V8 allows you to access diverse data and consolidate Web Services queries through a single SQL statement. DB2 still offers the broadest support for open standards.

DB2 Extenders

Only the DB2 XML Extender is included in V8.1 beta. The XML Extender provides data types to store XML documents in DB2 databases and new functions to work with these structured documents. These XML documents are stored as character data or external files and are managed by DB2. Retrieval functions enable you to retrieve complete documents or individual elements..

DB2 Administration Server for z/OS V8.1

IBM DB2 Universal Database V8.1 administration tools like the Control Center and other administration tools provide support to help you manage DB2 Databases on a number of platforms including OS/390 and z/OS server. The z/OS DB2 Administration Server (DAS) provides the functionality for running commands and making API calls on the OS/390 and z/OS servers.

z/OS OS/390 Enablement - CC390

This archive enables the DB2 Tools (including Control Center, Replication Center, Information Catalog Center and Development Center) , which use stored procedures, user-defined functions and batch programs to function properly in DB2 for OS/390 and z/OS servers.

DB2 V8.1 Processes

The process model used by all DB2 servers facilitates the communication that occurs between database servers and client and local applications. It also ensures that database applications are isolated from resources such as database control blocks and critical database files.

UNIX-based environments use an architecture based on **processes**. For example, the DB2 communications listeners are created as processes. Intel operating systems such as Windows NT use an architecture based on **threads** to maximize performance. For example, the DB2 communications listeners are created as threads within the DB2 server's system controller process. The process model for DB2 describes the behavior of processes and threads.

For each database being accessed, various processes/threads are started to deal with the various database tasks (for example, pre-fetching, communication, and logging).
Each process/thread of a client application has a single **coordinator agent** that operates on a database. A coordinator agent works on behalf of an application, and communicates to other agents using inter-process communications (IPC) or remote communication protocols.

DB2 architecture provides a **firewall** so that applications run in a different address space from DB2. The firewall protects the database and the database manager from applications, stored procedures, and user-defined functions (UDFs). A firewall maintains the integrity of the data in the databases, because an application programming error cannot overwrite an internal buffer or file of the database manager. It also improves reliability, because an application programming error cannot crash the database manager.

116.0   DB2 UDB ESE Version 8.1 – Detail

1. There is no more UDB EE or EEE there is now UDB ESE (Enterprise Server Edition).

2. Removed from the Run Time Client:-

    a. Command Center

3. Replaced at the Run Time Client:-

    a. Client Configuration Assistant

        By
    a. Configuration Assistant.

4. The Run Time Client is now administered by the Command Line Processor.

5. The DB2 OLAP Starter Kit, now not available in Version 8.

6. The Data Links Manager is not provided for in the DCE-DFS environment in Version 8.

7. The Development Center replaces the Stored Procedure Builder.

8. The Performance Configuration Wizard has been renamed the Configuration Advisor.

9. The Workload Performance Wizard has been renamed the Design Advisor.

10. A long table space is now called a large table space.

11. A nodegroup is now a database partition group.

12. 'Online Index Reorganization' is now referred to as 'Online Index de-fragmentation of leaf pages'.

13. A more comprehensive reorganization facility has been added for Version 8.

14. 'Country Code' has been changed to 'territory code'.

15. A Summary Table is now a 'Materialized Query Table'.

16. In V8, all Client Server data flows will use DRDA. Conversion tables for code pages will be installed on the Client.

17. IPX/SPX protocol is no longer supported.

18. SUNLINK SNA is no longer supported.

19. OS/2 is no longer supported.

20. PTX or NUMA-Q is not supported.

21. Windows 95 is no longer supported.

22. The General Replication Subscription function of the Satellite Administration Center is not supported.

23. The DB2Alert.log (alerts facility) has been removed. Users should use the Administration Notification Log as a replacement.

24. The Performance Monitor capability of the Control Center has been removed. This has been replaced by the Health Center and for multi-platforms the DB2 Performance Expert.

25. The type 3 JDBC driver (a.k.a. 'net driver' or 'applet driver' is going to "go away" soon. This means you should migrate existing DB2 JDBC applets to the new type 4 driver.

26. DCE Security has been removed. Customers are moving to Kerberos.

27. Light Weight Directory Access Protocol (LPAD) has become the industry standard. Customers should consider moving to LPAD as a replacement for DCE.

28. Version 8 removes the ability to authenticate a user at the DB2 Connect gateway. Authentication can now only be done at the Client or at the server using SEVER_ENCRYPT. These options must be cataloged at the Client in the database directory or left as NOT_ SPEC.

29. As you move your environment from V7 to V8, if you are in a situation where you migrate your Client machines to V8 before you migrate all your servers to V8 then there are several restrictions and limitations. These restrictions are not associated with DB2 Connect or the zSeries or OS/390 or the iSeries.

30. You can dynamically modify any parameter in the Database Configuration file without stopping and starting the Instance.

31. Through the VIEWS UNION ALL statement you can add and remove containers On-Line.

32. You can REORG and LOAD tables On-Line, the LOAD will lock the table at the table level, and not as the table space level as it used to do in earlier versions.
33. Type 2 indexes have been introduced as per OS/390 improving concurrency.

34. Version 8.1 works across the board, including AIX Version 5.1.

35. You may also use Version 8.1 with the P Series machines such as the P670 and P690.

36. You can create up to 32 physical partitions called LPARS in SMP and MPP machines. This is equivalent to working on mainframes with ADDRESS SPACES! Remember!!

117.0 UDFs

There are two types of User Defined Function (UDFs), which were introduced in DB2 V6. Sourced UDFs which are based on existing built-in DB2 functions and External UDFs which are pre-prepared pieces of code that accept parameters and return a value. External UDFs can be used to wrap data within application logic.

Once created, UDFs can be used anywhere a built-in function might be used. UDFs are split into two groups, deterministic and non-deterministic. Deterministic functions will always return the same value if provided with the same parameters e.g. 4 + 2 will always equal 6, and 8 – 2 will also always equal 6. However, non-deterministic functions may produce different results each time they are called, even if any of the supplied parameters are unchanged. An example of this is the IBM supplied UDF, MIDNIGHT SECONDS. MIDNIGHT SECONDS calculates and returns the number of seconds since midnight for the supplied argument. MIDNIGHT SECONDS will thus return a result which varies each time it is executed even though nothing has changed in the SQL statement.

UDFs can be "handy" because they enable the application to "ignore" physical data storage , an example of this might be formatting telephone numbers. Example:- a U.K. telephone is stored as 01753577733 and can be formatted into the U.K. telephone number format +44(1753) 577 773 by using a UDF.

There are many advantages of using UDFs, they allow more flexibility and are simple to use and transparent to the application. UDFs can be coded once and easily included in SQL. UDFs are reusable and easily updateable because there is only a single point of interchange.

Example 1:-
Here we use a sourced UDF which refers to the built in function AVG. The UDF is "sourced" on the existing built in function.

CREATE FUNCTION AVG_SALARY RETURNS (SALARY) SOURCE SYSIBM/ AVG(INTEGER)

Example 2:-
External UDF – Is functionality written in the application program, and defined to DB2 via the CREATE FUNCTION statement with reference to a load module.

CREATE FUNCTION CALC_BONUS (DECIMAL(6,2),DECIMAL(6,2) RETURNS DECIMAL(6,2) EXTERNAL NAME 'CBONUS' LANGUAGE COBOL;

PROGRAM:- CALC_BONUS (SALARY, COMMISSION);
            BONUS=(SALARY +COMMISSION)*.25;
            RETURNS;

Note:- There are two input parameters each DECIMAL (6,2) – salary and commission.

118.0 Understanding I/O

Avoiding I/O altogether is the best option for performance. The next best option is not to perform I/O to disk but to cache. Many factors are involved in optimizing and eliminating I/O operations. First, we must find out where and why I/O occurs by monitoring the subsystem and the application and then we must control the I/O by eliminating the I/O that adds unnecessary overheads to our applications.

DB2 uses bufferpools, hiperpools and group bufferpools to cache data and help minimize I/O to disk. Many disk storage controllers also provide and cache for intermediate data storage. These controllers can perform read-ahead operations if they detect that pages are being sequentially accessed, allowing a better cache hit ratio on the device.

There are different methods of bringing pages into the DB2 virtual buffer pools from disk. If the application needs a few pages or even just one, it performs a synchronous (normal) read. If the application is requesting several pages with sequential prefetch, the pages are read into the bufferpool in a single I/O, pre-staging them for read operations yet to come. This is known as a asynchronous read and often occurs concurrently with the application process. Dynamic prefetch can occur at execution time, which provides the same basics as sequential prefetch and gets invoked when a process called 'sequential detection' has determined that the pages are being accessed in sequential order. The last method is the list prefetch. This method operates with a RID (Row Identifier) list that is built from the index. The RIDs are sorted, and then pages can be accessed in order.

The pre-fetch operations are limited by the characteristics of the buffer pools. Be careful when defining the size of your buffer pools because you can inadvertently cause more I/O, because in the pre-fetch operation quantity is directly related to buffer pool size. "Normal" (the average of many sums") pre-fetch I/O (for 4K pages) is 32 pages per I/O. If your buffer pool is less than 1,000 pages, you may be getting only 8 or 16 pages per I/O, depending on the exact size, so you would be doing more I/O if your buffer pool is too small. Note that pre-fetch quantities vary by page size as well.

There are also various methods of writing data out to disk after it has been updated. The majority of writes from DB2 are done asynchronously, independently of the applications.
The other type of write is a synchronous write which causes the application to wait for the completion of the write. Synchronous writes can occur when exception conditions are encountered, such as when passing two DB2 checkpoints in a unit of work before an updated page is externalized, or when the virtual buffer pool that was holding the updated page is too small or it's immediate write threshold is reached.

Logging is also a source of many I/O operations, logging needs to be monitored and controlled. More about detailed logging operations at a later date.

119.0   NT Installation

Installation at the Server

The installation of DB2 for NT will come as a revelation to those used to the standard IBM installation routines. It is no longer a long winded, unwieldy process stuffed full of deeply searching questions and guaranteed to take up much precious time. Instead, it is just like that of any other NT program; you are guided by dialog boxes with helpful information and default settings.

So, where do you start? Strangely enough, you start by ensuring you have a user account with NT administrator privileges and a user name having eight characters or fewer.

Since the install process automatically registers the installer as a DB2 Systems Administrator (SYSADM), the NT name of that installer has to be restricted to eight characters.

I strongly recommend that you do create a new user and use this user only for administering DB2.

Now, login to the NT server as the user you have just created with SYSADM authority. You are going to have to re-boot the machine first, so make sure you are the only user on the machine. Close all programs that are running, including those that might have been auto started. Put the DB2 CD-ROM into the drive. The install program will now start up and you should see the 'WELCOME' dialog. The dialog screen simply tells you how wonderful DB2 is and how wonderful you are for using it. Click the <Next> button. A message announces that the "Space requirements are being calculated' after which you are asked to select the product or products you want to install. What you install will depend on what you have bought from IBM. No matter what version of DB2 UDB you choose to install, the Client Application Enabler (CAE) will be installed as well, whether you actively select it or not.

The next choice is the kind of installation you want to perform, the options are:-

- Typical
- Compact
- Custom

With each option comes an indication of the space required and a brief description. I recommend the selection of the "CUSTOM" option here, not because you will necessarily want to change anything, but simply because it allows you to "see" more of what is going on. In the next step, you should select everything that looks appropriate.

Rule of thumb:- If in doubt, select it.

This will save you much grief if you need to add the option at a later date.

When asked by the installation process whether to auto start DB2 and the Control Center, my advice is to reply "Yes" to both.

Choose a folder for DB2; the default suggestion of "DB2 for Windows NT" is a good one.

Next comes the question of communications protocol that you can configure. For the DB2 instance and Administration Server. Click on the Customize button and DB2 will show you what it intends to use as defaults. In my experience these are fine. However if you use TCP/IP you should select the DB2 Instance Customize and, in the Properties for TCP/IP make a note of the port number, usually 50000, because this may prove useful later.

You are now asked for a username and password word to enable the Administration Server to log on to the system. The user must have administrator privileges, so unless you have reasons for doing otherwise, enter the details of the user name and password that you are presently using for this installation.

You are then given the opportunity to review the decisions made so far, and if there is anything you wish to change you can do this at this point. Once you are happy, click on the <install> button and the installation should proceed smoothly.

Finally you are asked to re-boot the machine, but remember to take out the CD first.

You have now installed DB2 UDB on the NT Server.

120.0 The Sample Database

After the machine re-boots it should auto start the Command Center and DB2 First Steps.

DB2 First Steps is a brief guide to DB2. It is worth investigating because it introduces you to DB2's excellent help system which includes On-line manuals. First Steps is also the place from which you can 'Create the Sample Database'. I very strongly recommend that you do create the Sample Database. SAMPLE takes up about 12 MB of disk space, and can be used as a 'sand box' by all users. It is also referred to in IBM's documentation. Creating the SAMPLE database takes very little time. It should take no more than two minutes on a 200 MHz machine with 100 MB of memory.

What you should find if all has gone well

Once DB2 and the SAMPLE database have been installed it is probably worth summarizing the changes that are made to your server with DB2 installed and what you should now find running on it. That way, if for any reason the installation fails, you can compare what you actually have with what you should have in an effort to track down the likely cause of failure.

The installation of DB2 creates two directories off the root:-

C:\Sqllib and C:\Db2log
DB2 itself is located under a maze of sub-folders that lurk under Sqllib. A Custom installation should take up in the region of 100 MB of disk space.

All that Db2log contains is a log file called, quite reasonably, db2.log.

DB2.log, logs records as ASCII text, what the install program should have done is worth consulting if anything did go wrong an error message should be in this file.

Next, if you check <Services> (<Start>, <Settings>, <Control Panel>, <Services>) you should find that two new services are running.

- DB2 – DB2
- DB2 – DB2DAS00

DB2 is the database engine and DB2DAS00 is the administration instance (logical server) that administers DB2 and enables remote administration. These are both listed for automatic start up.

A third service, called DB2 Security Server is by default not started and is left for manual start up. But it is better to set this for automatic start up unless you are sure you don't need it. To select automatic startup, double click on the name of the service and select automatic. It will start all by itself the next time the machine is restarted.

If you find the system is not working after installation, and you can not find any of these folders, try re-running the installation program.

121.0 XML Extender

DB2 XML Extender. DB2 UDB version 7.1 and up includes the XML Extender, which allows you to store XML documents as a new data type. The XML Extender lets you decompose, store, index, and search against XML component parts. You can retrieve either the entire XML document or its component parts. XML documents can be stored directly in DB2 as character data or stored as external files under DB2's control. XML Extender functionality, combined with enhancements to Net.Data, provide a simpler way of exchanging and storing documents electronically. **Net.Data,** available free of charge with DB2 UDB, is a simple but very powerful tool for connecting Web applications to DB2. Version 7.1 enhancements include features that provide XML tag generation from macros and the ability to specify an XML style sheet for formatting and displaying XML reports.

NEW BUSINESS INTELLIGENCE SUPPORT

Two new data warehouse management features and analysis enhancements in version 7.1, an OLAP starter kit, and an optional Spatial Extender promise to simplify and speed business intelligence solution deployment.

Integrated Data Warehouse Center. Version 7.1 introduces the integrated Data Warehouse Center, a new warehouse management function built into DB2 Control Center. Based on Visual Warehouse, Data Warehouse Center lets users define data sources, manage data movement, populate a metadata repository, create generic schema models, and more. Data sources can include DB2 sources, other relational databases, and flat file sources. Data Warehouse Center also allows you to design logical star schemas for the new OLAP Server Starter Kit.

DB2 Warehouse Manager, a separate product available for an additional charge, includes additional warehouse agents, DB2 Query Patroller, the Information Catalog Manager, and QMF for Windows.

DB2 Query Patroller, introduced in version 6.1, is now available for the Enterprise Edition and the Enterprise-Extended Edition. Later this year, the product will also be available to DB2 HP-UX and Sequent users. In earlier releases, DB2 Query Patroller was a stand-alone, optional product. Now DB2 Query Patroller, which helps manage, govern, and schedule business queries to make the most of hardware resources during peak and off-peak periods, is integrated into DB2 Warehouse Manager. Whereas DB2 Query Patroller could previously manage only dynamic SQL coming through ODBC clients, a new integrated trap for client code now enables DB2 Query Patroller to manage dynamic queries regardless of their originating source, including command line processor SQL. IBM often tells clients that DB2 UDB is a kind, gentle, and loving database that will go to great effort to allow a SQL query to complete successfully. Now, a new DB2 Query Patroller retry option provides the capability for queries to be automatically resubmitted for execution depending on the original cause of failure. The combined odds of users getting the timely results they need are now even greater.

John V. McLean

The Information Catalog Manager, previously known as DataGuide, includes enhancements that help users find and access information by populating the catalog with information from the Data Warehouse Center and analytical tools (such as DB2 OLAP Server, Lotus 1-2-3, Hyperion Essbase Server, and Microsoft Excel). Information Catalog Manager also lets users register shared information, searches available information objects to find relevant information, and displays object metadata.

122.0 Reduced lock contentions through new lock semantics

For some access paths, locks may be acquired on rows that don't qualify the statement's predicates because of locking semantics which in which DB2 evaluates predicates on committed data only. Therefore, a requested lock could get suspended waiting for a row that does not qualify the predicates. Some applications such as SAP do not need this semantic.

A new system parameter EVALUNC, controls whether the old, pre V7, semantics or the new ones will be used. The default value NO specifies the traditional semantics. A word of caution:- You have to fully understand this feature's ramifications to be sure that the application semantics accepts them.

123.0 Faster inserts through asynchronous preformatting

Prior to V7, mass insert processing was affected by space preformatting. Namely, when the formatted space was exhausted, a synchronous task for formatting the additional space was scheduled. Inserts must wait until this task completes it's processing. During heavy insert processing these situations can appear often and can significantly slow down the overall process. With DB2 V7 the preformatting is an asynchronous process that gets triggered before the format space is exhausted. When a new page for insert is close to the end of formatted pages and there is more allocated unformatted space in the data set, an asynchronous format task will format the next range of pages. This will ensure that an insert rarely has to wait for the page to be formatted. Additionally, the space search performance is improved because the search will not happen at each logical extend but only at each physical extend. This feature is internal to DB2 and no user specification is necessary.

124.0 Faster updates through improved log write performance

The log latching in V7 and up is reduced, resulting in the ability to support higher log write rates. Also, the amount of data logged for the variable length columns has been reduced by logging the changed row from the byte of the first changed column to the last byte of the last changed column. This feature does not apply for compressed tablespaces.

125.0 Row-level locking for selected catalog tables

Earlier versions of DB2 did not allow ALTER LOCKSIZE against the catalog tablespaces. In most cases the catalog tablespaces use page level locking, and that can create lock contention, especially during parallel DDL executions. In V7, ALTER TABLESPACE LOCKSISE is supported for tablespaces of the DB2 catalog that do not contain links – all tablespaces except SYSDBASE, SYSPLAN, SYSVIEWS, SYSDBAUT and SYSGROUP. Attempts to ALTER TABLESPACE LOCKSIZE for one of these tablespaces will result in a SQL return code of –607.

126.0 Faster query execution through FETCH FIRST n ROWS

The new FETCH FIRST n ROWS option speeds up fetch processing in cases in which the number of rows that need to be fetched is known in advance. Another interesting use is to check if a table is empty. Use the following query:-

SELECT 1 FROM TAB FETCH FIRST 1 ROW

Unless explicitly specified, OPTIMIZE FOR n ROWS is implied.

127.0 Faster Warehouse queries via UNION in views

Using UNION in views streamlines warehouse processing for better performance.

128.0 The DB2 Performance Configuration Wizard

The DB2 Performance Configuration Wizard, is an expert tool for the configuration of DB2 Universal Databases. This utility has shown dramatic results for tuning and configuring DB2 servers on UNIX and Windows platforms, particularly on OLTP (transactional) systems. The recognition of the essential need for administration and design tools has spurred renewed interest among leading RDBMS vendors. The recent proliferation of papers on index and materialized view (summary table) selection, and the development of industrial applications in this regard by leading vendors, such as Microsoft, IBM, and Oracle, as well as numerous third party administration tools vendors, are all testament to the growing corporate recognition of this important area of investigation. The DB2 Performance Configuration Wizard is a key feature in DB2's self-managing technology portfolio.

129.0 REORG timeout and drain retry options

Before V7, the timeout value for a REORG was at least equal to the value specified as the transaction's timeout. It was calculated as a product of the system parameters IRLMRWT (the transactions timeout value) and UTIMOUT, the utility timeout factor (greater than or equal to 1) that allows the REORG to wait longer for a resource than a regular statement would. This behavior is not really what customers want. It's often more appropriate if REORG times out before any user transaction. In V7, the new REORG option DRAIN_WAIT (with possible values of 1 to 1,800 seconds) specifies how long the REORG will wait to acquire drains. Draining is a mechanism used to take over an object and serialize access to it. The new option applies to drains in the UNLOAD, LOG and SWITCH phases of the REORG. When multiple objects need draining (for example, a tablespace and it's index), the DRAIN_ WAIT value is the aggregate time. In contrast, the timeout value is per object in Version 6 or lower. Use the DRAIN_WAIT option in conjunction with the new retry logic that preserves REORG processing done before DRAIN_WAIT is exhausted, otherwise, after a timeout, the REORG process would terminate without accomplishing anything.

130.0 Introduction to Version 8.1 ESE (a.k.a EEE)

This section attempts to detail installation and partitioning options for DB2 UDB EEE V8.1 (a.k.a. ESE). At the end of the document there are recommendations. Whether these techniques are adopted or not, the document lays out what is involved in manual processes even though automatic alternatives may be chosen.

130.0.1 How to Choose an Installation Method

Let us review three installation methods and discuss the advantages and disadvantages of each:-

Using the DB2 Installer program

This is the simplest method because the DB2 Installer is a menu driven program provided with the product. It can be used to configure some of the DB2 communication environment and to create DB2 and DAS (see below) instances and the DB2 instance owner user and group.

However, you will have to repeat the installation on every host. If your configuration is eight hosts or less (relatively small) you could establish a session on each host and perform the installation this way. If your DB2 Cluster has a large number of hosts, this could be a time-consuming process.

Using SMIT

You must manually create the DB2 instance and the DAS instance and configure all DB2 communications after installation using SMIT.

Using SMIT, you can save the INSTALLP command used to install UDB EEE. You can then execute this INSTALLP command on the other hosts, either from a separate session for each host, or using the **dsh** program to automate the process.

Using the INSTALLP command under dsh

The **dsh** program executes an application on all hosts in a defined environment. This is probably the easiest method to use if you have many hosts in your DB2 cluster. You must still manually create the DB2 and DAS instance, the fenced user and configure all DB2 communication. You must also check the installation was successful on each host.

130.0.2 Installing and Configuring DB2 UDB EEE

Below are the steps necessary for a manual install on four SP Nodes. The various names selected are completely arbitrary.

To manually install and configure DB2 UDB EEE perform the following tasks:-

- Make file system for the DB2 Instance's home directory (home/tp3an01) available across NFS to all the SP Nodes.

- Install DB2 UDB EEE.

- Add a Group and a User for the DB2 Instance.

- Add entries to /etc/services/ on all SP Nodes.

- Create the DBB2 Instance.

- Edit the **.**profile file

- Edit the db2nodes.cfg file

- Edit the .rhosts (remote hosts) file.

- Set up db2diag.log and syslog.

- Start the DB2 Instance.

- Create DB2 Objects.

8. <u>Make the file system for the DB2 instance's home directory</u>
    (home/SP Node name) across NFS to all SP Nodes.

    Smitty nfs

    → Network File sytem (NFS)
    -→ Add a directory to the Exports List

    PATHNAME of directory to export   /home/tp3an01
    HOSTS allowed root access   tp3an05, tp3an09, tp3an13

9. Install DB2 UDB EEE

Log in as root on the first SP node xxxxxxxx
Change directory into the directory where the DB2 images are located
Run smitty installp then specify . (dot) as the input directory.
Select F4 against Software to install.
Choose License for DB2 UDB EEE
Edit smit.log to find installp command that was run.

10. Add a Group and a User for the DB2 Instance

Smitty group
→ Add a Group

Group Name ……………… dg2asgrp

To add a User:-

Smitty User
 → Add a User

User Name…………………….. db2inst1
Primary Group………………… dn2asgrp
Home Directory ……………… /home/tp3an01/dn2inst1

Before distributing this user to the SP nodes we need to:

Set it's password because to logon as a new user, a new password must exist.
Change it's password because by default when we first log in as this new user
the system will prompt us to change the password.

11. Adding Services Entries

Before creating the DB2 UBD EEE instance entries must be made in the /etc/services/
file on all SP nodes to be included in the instance. If we are using 4 database partitions
(DPs) per SP node, we need to reserve 4 ports per SP node. To support HACMP
failover, another 4 ports must be reserved making a total of 8.

Add these lines to /etc/services:

DB2_db2inst1   3000/tcp
DB2_db2inst_END   3007/tcp

This means that ports 3000 to 3007 inclusive are reserved for DB2

12. Now we are ready to create the instance

Log in as root at tp3an01

```
tp3an01[/]> cd /usr/lpp/db2_08_00/instance
tp3an01[/usr/lpp/db2_08_00/instance]> ./db2icrt –ud2inst1 db2inst1
DBI1070 Program db2icrtg completed successfully
```

13. Add Profile Entries

Entries need to be added to .profile of db2inst so that environmental variables and DB2 profile variables are set. Add these lines to:-
/home/tp3an01/db2inst1/.profile:-

```
. ~/sqllib/db2profile
# Set the default DB
db2set –I db2inst1 db2dbdft=tcpd30
```

14. Edit the db2nodes.cfg file

The file db2nodes.cfg in the directory $INSTHOME/sqllib defines which database partition servers will be started when a db2start is issued.
Using vi add these lines:-

```
1 tp3an01 0 tp3sn01
2 tp3an01 1 tp3sn01
3 tp3an01 2 tp3sn01
4 tp3an01 3 tp3sn01
5 tp3an05 0 tp3sn01
6 tp3an05 1 tp3sn05
7 tp3an05 2 tp3sn05
8 tp3an05 3 tp3sn05
```

Start numbering the DP (database partitions –nodenums column one) at 1 so they would be synchronized with the host names of the SP nodes.

Hostname (2nd column) is defined as the first Ethernet interface (en0) on each SP node.

Logical port (3rd column) must be 0, 1, 2, 3 for the four DPs on each SP node

Switchname (4th. column) is the network interface of the switch on each SP node.

8. Creating .rhosts Entries

Before issuing db2start we have to make sure that remote execution permission is defined across the SP nodes. In other words, if we execute rsp tp3an05 date from tp3an01 we will not be prompted for a password. This needs to be true for all the network interfaces defined in the db2nodes.cfg file.

Set permission by adding these lines to .rhosts in /home/tp3an01/db2inst1 on tp3an001:-

```
tp3an01   db2inst1
tp3sn01   db2inst1
tp3an05   db2inst1
tp3sn05   db2inst1
```

9. Set up the db2diag.log

Most DB2 informational and error messages are written to db2daig.log. This file is by default stored in the $INSTHOME/sqllib/db2dump directory, which is NFS shared to all the SP nodes. It is a good idea to change this path so that the db2diag.log is written locally on each SP node.

To do this:- db2 update dbm cfg using diagpath /tmp/db2

Where /tmp/db2 is the file system that exists locally on each SP node.

10. Setting up syslog.conf

Some DB2 informational and error messages apply to the SP node and are independent of the DB2 instance. These messages are captured by configuring the AIX syslog daemon:-

Add to /etc/syslog.conf:- user.warn /tmp/db2/syslog.db2

Create the /tmp/db2/syslog.db2 file:- touch tmp/db2/syslog.db2

Find the process id (pid) of the syslog process

ps –ef|grep syslog

Send a –1 signal to this process using Kill:-

Kill –1 <syslogd-pid>

Run this set of commands on each SP node.

11. <u>Start the instance</u>

Now we are ready to start the instance:-

Db2start

You can start a single DP server:- db2start nodenum 16

12. <u>Create the Database, Nodegroups, Table Spaces and Tables</u>

Db2 terminate
…. Various Creates here.

130.0.3   Configuring DB2 Communications

There are a number of tasks to complete to enable communication between hosts in your DB2 cluster. There follows a check list of these tasks:-

1.   Make sure all hosts communicate with each other through TCP/IP. From each host, ensure that the host names of the other nodes can be resolved, through the /etc/services file.

2.   Reserve ports for the DB2 Instance in the etc/services file.

3.   Make sure the DB2 instance owner can execute commands using the **rsh** command on all hosts to be used in the DB2 cluster.

130.0.4   The db2nodes.cfg Configuration File

The db2nodes.cfg file contains information about the database partitions in a DB2 instance. The file must be placed in the sqllib directory in the DB2 instance owner's home directory ($INSTHOME/sqllib). There is one file for every DB2 instance on your system. The file contains one entry for each database partition for that particular DB2 instance.

Before you create a DB2 instance you must decide how many database partitions will be defined per host so that you can reserve the correct number of ports in the etc/services file. You should complete the db2nodes.cfg file after the DB2 instance creation.

You can edit the db2nodes.cfg file with any test editor. The file is locked after you Issue the db2start command. DB2STOP and DB2START allocates the partitions as Set in the DB2NODES.CFG file.

There are four fields in the db2nodes.cfg file:-

1. Nodenum
2. Host Name
3. Logical Port
4. Netname (siwtch).

The db2nodes.cfg file looks like:-

```
1   host1   0   switch1
2   host2   1   switch2
3   host3   2   switch3
4   host4   3   switch4
```

The /etc/services looks like:-

```
DB2_test        500000/tcp
DB2_test_END  500003/tcp
```

130.0.5   Enabling the Execution of Remote Commands

In a UDB EEE instance, each host must have authority to perform remote commands on the other hosts that make up the instance. This is done by a .rhosts file in the home directory of the instance owner. The home directory of the instance owner is NFS mounted on all the owner hosts. Therefore a single copy of the .rhosts file is accessible from all the hosts. Alternatively, you can create a /etc/hosts.equiv file on every host in the instance. However, this is more difficult to maintain.

The .rhosts file looks like:-

```
Host Name        Userid

Fo1no5           db2inst2
F01no6           db2inst2
F01no7           db2inst2
```

In the example above, all users that can issue remote commands on a given host(s) are listed. Make sure the IP addresses of the hosts can be resolved using the /etc/hosts file.

130.0.6 Creating a DB2 Instance in EEE

You create and store databases in DB2 instances.
(also see Instance description below).

This is what is involved in creating an instance in DB2:-

1. The same User with the same User Id., Group and Group Id.,
   must exist on all your hosts in your DB2 cluster.

2. The home directory of the DB2 instance owner should exist on one host and be
   NFS mounted on all remaining hosts in the cluster.

3. Create the instance on only one DB2 host. This host should be the host where
   the home directory of the DB2 instance exists. Even if your DB2 cluster involves
   multiple database partitions on the same host, you create the DB2 instance only
   once on one host. You can create the DB2 instance by one of the following
   methods:-

   a. Using the DB2 Installer program,

   or

   b. Using the **db2icrt** command.

   If you use the DB2 Installer program to create a DB2 instance, the utility will
   also configure the host for client/server communication (this is not the same as
   configuring communication between database partitions).

130.0.7 DB2 instance, /etc/services, and db2nodes.cfg file

Remember that the entries for communicating between database partitions in the /etc/
services file must be done **before** the DB2 instance is created, or the DB2 instance will
fail. Once the instance is successfully created you should configure the db2nodes.cfg file
for your environment.

130.0.8 Partitioning Overview

Partitioning maps form an important part of the process used by the database manager to
allocate the rows of a table across the database partitions in a database. To understand why
DB2 UDB EEE uses partitioning maps, we must first understand this process:-

DB2 UDB EEE uses a hashing algorithm to assign a given row of a table to a corresponding
database partition. When a table is created, a partitioning key is defined for that table. This

partitioning key consists of one or many columns. The database manager uses the value of the partitioning key for a given row as the input to the hashing algorithm which is part of the LOAD function. The output of the hashing algorithm is a number between 0 and 4095 which is used as a displacement into the partitioning map. I use the term "index" for this displacement. So if you have eight partitions they will be divided into 512 equal pieces (4096 / 8 = 512). If the displacement calculated is 1540 then the row will be stored in the fourth partition (512 * 3 = 1536) + 4 = 1540 = partition number 4.

A partitioning map is a sequential list (or Vector) of 4096 database partition numbers. Each partitioning map corresponds to only one nodegroup and is created when the nodegroup is created or when data is redistributed.

Often the partitioning map is represented as being formed of two rows. The first row represents the index which ranges from 0 to 4095 and the and the second row is the value at that index which is the database partition number.

A default partitioning map is created when a node group is created. The map is stored in the SYSIBM.SYSPARTITIONMAPS SYSTEM Catalog Table. The default partitioning map contains database partition numbers of the nodegroup assigned in a "round-robin" fashion. It is also possible to use partitioning maps which do not use this round-robin system. These maps are known as customized partitioning maps.

There is a utility called db2gmap that can extract the partitioning map of a Nodegroup from the SYSIBM.SYSPARTITIONMAPS system catalog table.

For example, if you have a nodegroup called NG123 that consists of database Partitions 1, 2 and 3 in a database called SAMPLE and you want to get the partitioning map of this nodegroup to be written to a file called NG123map.txt then issue the following command:-

db2gmap –d sample –g NG123 –m NG123map.txt

131.0   Instance Description

A UDB server provides database services to applications running on it's own machine and also on remote client machines. Multiple copies of a UDB server can run on the same physical machine. Each copy of a UDB server is called an *Instance* and has it's own name, this is called the *Instance Name*. Although the instances on a given machine share code, they operate independently of each other. Each instance can have it's own System Administrator and it's own set of configuration parameters. For example, one instance may be configured for high volume transaction processing, while another may be configured for decision support applications, or one instance may be used for Development, another for Quality Assurance and yet another for Production.

When ever a UDB server is installed, two instances are created. One of these typically named DB2DAS00, is an administration server used by UDB itself to manage it's own
internal affairs. The other typically named DB2 is created to manage User databases.
After the initial installation of UDB additional instances may be created by using the command *db2icrt*.

Each UDB instance can create and manage one or more databases. A database is a name under which there are a collection of physical objects such as table spaces, tables, indexes etc, etc. A database itself has no physical properties.

Each database belongs to a particular instance and resides on the machine where that instance is installed. Each database has a name which is chosen at the time the database is created. Each database has (is supplied with) a set of System Catalog Tables, these tables are automatically maintained by the system. These System Catalog Tables contain information about all the objects (table spaces, tables, indexes etc.) that are stored in the database, and about users of the database and their access privileges. Information about the database can be retrieved from the System Catalog Tables using SQL queries.

Each UDB *Client* (usually a User's terminal) maintains a list of instances and databases with which it knows how to communicate. Installed with each UDB Client is a graphical tool called the CCA (Client Configuration Assistant) that can be used to add new instances and databases to the list.

Before interacting with a database, a UDB User or application program must establish a connection with that database. A connection is established with the CONNECT command which checks to make sure that the User or application is authorized to access the given database. Under certain circumstances it is possible for the User or application to be connected to multiple databases at the same time.

Some UDB commands operate at the level of an instance rather than at the database level. For example, a command might change the configuration parameters of an instance, or create a new database to be managed by a given instance. Commands executed on a server machine are directed to the instance named in the environment variable DB2INSTANCE. On a Client machine, an ATTACH command can be used to name the instance to which the instance-level commands are to be directed.

132.0 Creating the DAS Instance
Manual Option

The DB2 Administration Server (DAS) is a special DB2 Instance used for managing DB2 Servers. The DB2 Administration Server instance on a DB2 server machine provides a remote Client the ability to administer and detect the instances and databases at the server.

You can think of DAS as being global to a machine. The environment of the DAS instance is separate and different from that of the DB2 Instance Owner. In a EEE environment you should first configure the DAS instances. In some set ups this is done automatically.

133.0   Create an AIX Group and User for DAS Instance Owner

Create an AIX group and user for the DAS Instance Owner. Use the same method when creating the AIX Group and User for the DB2 Instance Owner.

```
>$ mkgroup id=nnn db2asgrp2
>$ mkuser id=nnnn pgrp=db2asgp2  groups=db2asgp2
home = /home/db2as2 db2as2
>$ passwd db2as2
Changing password for "db2as2"
Db2as2's New Password
Enter New Password again
>$
>$ login db2as2
db2as2's Password:
3004-610 You  are required to change your password please choose one
db2as2's New Password
Enter the new Password again.
>$ /home/db2as2 > exit
```

The mkgroup command creates a group named db2asgp2 with GID=nnn

The mkuser command creates a user named db2as2 with UID = nnnn
who belongs to group db2asgp2 and has a home directory of /home/db2as2.

The home directory (/home/db2as2) is created on the Control workstation.

The **passwd** command lets you create an initial password for the user db2as2.

The login command is used to stop the system from asking for password again at the next login.

134.0 Intra-Partition Parallelism

One of the most important concepts of UDB is UDB parallelism and the concept of a partition.

In a parallel system each database can be split into several parts called partitions. Each table in the database can have some of its rows in each partition. It is helpful to visualize each partition as running on a separate machine, although it is possible for more that one partition to be assigned to the same machine.

Each database partition has it's own log and it's own set of indexes.
Intra-partition parallelism refers to simultaneous processes within a single partition.
Inter-partition parallelism refers to simultaneous processes in multiple partitions.
These two kinds of parallelism are independent of each other, and their relative importance depends on your hardware configuration.

IntraPP is often used on a symmetric multiprocessor (SMP) machine, in which multiple processors share common memory and disks, To exploit IntraPP, the optimizer generates access plans that contain multiple threads that can be active simultaneously during processing of an SQL statement.

Since intraPP takes place within a single database partition, all the threads have access to all the data in the partition.

The number of threads in an access plan, called the "degree" of the plan, may be more or less than the number of physical processors on the system.

To use intraPP with degree 4, the optimizer may generate a plan with four threads each scans a different part of the table.

Even if four threads are shared by two or three processors, a significant performance improvement can result.

IntraPP is completely transparent to SQL it places no limitations on SQL statements.

To exploit intraPP set the database manager configuration parameter INTRA_PARALLEL to YES. The optimizer will then generate multiple threads and the run time system will execute these plans using multiple processors.

BIND with DEGREE ANY – no limit on the degree chosen by the optimizer. Default is when configuration DFT_DEGREE is 1. CURRENT_DEGREE can be set to ANY default is 1 as per the DFT_DEGREE parameter.

MAX_QUERYDEGREE and SET RUNTIME DEGREE also limit the number of threads running in a plan.

135.0 What type of partitioning for most Installations.

Inter-partition parallelism – refers to the ability to break up a query into multiple parts across multiple partitions of a partitioned database, on one machine or multiple machines. The query is run in parallel.

Simultaneous Intra-partition and Inter-partition parallelism – You can use Intra and Inter partition parallelism together and at the same time, giving you the best of both worlds.

This combination gives you two dimensions of parallelism, resulting in a dramatic increase in performance and an increase n the speed at which queries are processed.

136.0   Connections Foreword

DB2 Connect gives clients on the LAN access to data stored on host systems. It provides applications with transparent access to host data through a standard architecture for managing distributed data. This standard known as Distributed Relation Database Architecture (DRDA) allows applications to establish a fast connection to databases on MVS, OS/390, AS400, VM, VSE, AIX (V6.2) and NT hosts.

Applications that run on any of the supported platforms can work with data transparently, as if a local database server managed it.

The following tools and products can access host data easily by using DB2 Connect:-

Lotus 123 and Microsoft Excel to analyze real-time data without the cost and complexity of data extract procedures and import procedures.

Decision support tools such as Business Objects, Intersolv Database Editor and Crystal Reports to provide real-time information.

Personal database products such as Lotus Approach and Microsoft Access.

Development tools such as IBM's Visual Age, PowerBuilder, Microsoft's Visual Basis and Borland's Delphi to create Client/Server solutions.

With DB2 Connect multiple clients can connect to host data and can significantly reduce the effort to establish and maintain access to enterprise data. Also, the user can use IBM's data replication tools to propagate data among numerous databases (see my paper on 'Replication' 6/22/99) this allows (amongst other things) access to databases for informational or prototyping purposes without interfering with operational systems.

137.0 Managing Connections to Databases

The <u>C</u>lient <u>C</u>onfiguration <u>A</u>ssistant helps the user manage database connections to remote database servers. It leads the user through the necessary steps to configure and manage DB2 clients, while at the same time hides many of the steps required by automation. With the CCA the user can do as follows:-

Define database connections in three ways so that applications can use the databases:-

1. Search the network for available databases and select one. Client access is automatically set up for that database.

2. Use a database access profile provided by the database administrator to automatically define connections.

3. Manually configure a connection to a database by entering a few required parameters.

Remove cataloged databases or modify the properties of a cataloged database.

Test connections to local or remote databases identified on the user's system to ensure the user can connect to the server required.

Bind applications to a database by selecting utilities or bind files from a list.

Tune the client configuration interface parameters.

Set up a connection to a DRDA (covered in detail later in this document) server if DB2 Connect is installed.

Export the values used for the user to another client, so that the setup can be duplicated from one client to another.

Register the database as an ODBC data source and customize the settings for the ODBC application that the user is using.

138.0   Key Tool

One of the more useful tools available for remote connectivity is the DB2 Software Developer's Kit. The DB2 SDK is a collection of tools designed to meet the needs of database application developers. It includes libraries, header files, documented API's and sample programs to build text based or multimedia or object oriented applications. A platform specific version of DB2 SDK is available for each supported operating system.

The DB2 SDK allows the user to develop applications that use the following interfaces:-

a.	Embedded SQL
b.	Call Level Interface (CLI) compatible with Microsoft ODBC drivers
c.	Java Database Connectivity (JDBC)
d.	DB2 Application Programming Interfaces (APIs) to access database utilities
e.	Can access other database servers that support DRDA
f.	Supports several programming languages including COBOL and C++

139.0   DRDA

When discussing distributed DB2 data, it is necessary to cover DRDA (Distributed Relational Database Architecture) – an architecture developed by IBM that enables relational data to be distributed among multiple platforms. Both like and unlike platforms can communicate with each other. The platforms do not need to be the same, as long as they both conform to the DRDA specifications, they can communicate.
DRDA can be considered a sort of universal distributed protocol.

DRDA is made up of a set of rules (protocols) that enable a user to access distributed data regardless of where it physically resides. It provides an open, robust, distributed environment and methods of coordinating communication among distributed locations.

A distinction should be made between the architecture and the implementation of DRDA. DRDA describes the architecture for distributed data processing and nothing more. DRDA defines the rules for accessing the distributed data, but does not provide the actual Application Programming Interfaces (APIs) to perform the access. Thus, DRDA is not an actual program, rather, it is similar to the specifications for a program.

When a DBMS is said to be DRDA-Compliant then it follows DRDA Specifications. DB2 is a DRDA-Compliant RDBMS product.

Benefits of using DRDA

DRDA is only one protocol for supporting distributed RDBMS. Of course, if the installation is a DB2 user then DRDA is probably the only protocol for distributed data that matters. The biggest benefit provided by DRDA is a clearly stated set of rules for supporting distributed data access.

Any product that follow the DRDA rules can seamlessly integrate with any other DRDA-Compliant product. Furthermore, DRDA-Compliance RDBMSs support full data distribution, including multi-site update. The biggest advantage however is that DRDA is available today, and an ever increasing number of vendors are jumping on the DRDA-Compliance 'band wagon'.

Alternatives to DRDA

An alternative to using DRDA is to utilize a 'gateway' product to access distributed data. Gateways are Comprised of at least two components – one for each distributed location. These parts communicate with one another. With DB2 a host based gateway component is necessary, it functions as another DB2 application. Most 'gateway' products typically support Dynamic SQL only.

140.0   DRDA Functions

Three functions are used by DRDA to provide distributed relational data access:-

1. Application Requester (AR)
2. Application Server (AS)
3. Database Server (DS)

These three inter-operate to enable distributed data access.

<u>DRDA Application Requester (AR)</u>

This function of DRDA enables SQL and program preparation requests to be requested by the application Programs. The AR accepts SQL requests from an application and sends them to the appropriate Application Server(s) (ASs) for subsequent processing.

<u>DRDA Application Server (AS)</u>

The DRDA application server (AS) function receives requests from application requesters and processes them. These requests can be either SQL statements or program preparation requests. The AS acts upon the portions that can be processed and forwards the remainder to DRDA database servers for subsequent processing. The AR is connected to the AS using a communication protocol called the Application Support Protocol (ASP). The ASP is responsible for providing the appropriate level of data conversion (example:- ASCII to EBCDIC or EBCDIC to ASCII).

<u>DRDA Database Server (DS)</u>

The DRDA database server (DS) receives requests from application servers or other database servers, these requests can be either SQL statements or program preparation requests. The database server processes what it can and then forwards the remainder of work to another database server.

Note:- A DS request may be for a component of an SQL statement, this would occur if data is distributed across two different sub-systems and a JOIN is requested from tables at two different locations. The DS uses the Database Support Protocol to connect an Application Server to a Database Server.

141.0   DATAWAREHOUSE Revisited

<u>Data Profiling and Analysis Solution</u>

One of the greatest obstacles to the successful and timely completion of enterprise data management and integration has been the need for a thorough source system analysis deriving and mapping the true data content of source and legacy systems as a first step towards successful integration of that data.

Until recently, data profiling has been a manual, labor-intensive, resource-devouring and error-prone requirement for almost every major data integration project undertaken by any enterprise.

Ascential Software best-of-class, automated data profiling and source system analysis solution, automates this all-important first step in data integration dramatically reducing the time it takes to profile data from months to weeks or even days. MetaRecon also drastically reduces the overall time to complete large-scale data integration projects, and can automatically create the ETL job definitions you can execute with Ascentials DataStage product. This even further reduces downstream development time.

The result is a faster, bigger ROI for enterprise applications like CRM, SCM, ERP, BI, e-business and data warehousing.

In broad terms, data profiling is the process of collecting all pertinent information on he content and structure of an existing set of data sources and examining that data to provide a correct and complete model for a target database. A complete data profile contains the essential detailed information on data values, attributes, dependencies and relationships.

Value to your business

Reduces time needed to analyze source systems by 90%
Ensures correct data before starting development
Cuts development time by automated ETL job definition
 Reduces costs by catching problems during design stage instead of production

Capabilities

Automated profiling and analysis of source system data content, structure, quality and dependencies
Optimized target database definition (DDL) generation
Automatic DataStage ETL job generation
Source-to-target mapping environment
Single, open repository for storing the analysis results and metadata
Wide range of pre-built reports
 Multiple clients, multiple servers, a variety of supported platforms

The data profile forms the basis for developing accurate source to target transformation maps during the subsequent mapping steps. Data Mapping is the process of matching the source data profile to either an existing target database design or using it to create to create a target database design. Data Mapping needs to be performed at the entity, attribute and at times data value level. This information in turn is used in conjunction with third party data migration tools to extract, scrub, transform and load the data from the old systems.

Digging out the essential characteristics of older systems can only be accomplished in an accurate way through analysis of the actual data, and much of that digging can only be completes through automated processes. This is the strength of data profiling and mapping software.

The up front data profiling and mapping phases are the most difficult parts. Approximately 60% to 80% of the project time is spent on these phases (understanding the legacy data sources and creating an accurate data model and set of mapping specifications). A successful outcome of a data migration project (in terms of project completion time, cost and quality of results) depends on this part being done correctly.
Data profiling and mapping software provides utility to organizations faced with analyzing existing data sources as the basis for migration, re-emgineering or consolidation.

Data Warehousing

The idea behind data warehousing is one that has been performed by IT professionals for years.

It enables end users to have access to corporate operational data to follow and respond to business trends.

The true benefit of data warehousing lies not with the conceptual components embodying the data warehouse, but the combination of these concepts into a single, unified implementation that is novel and worthwhile.

Consider the typical DP shop. Data is stored in many locations, in many different formats, and is managed by many different DBMSs from multiple vendors. It is difficult if not impossible to access and use data in this environment without a consistent blueprint from which to operate. The blue print is the data warehouse.

Data warehousing enables the organization to make information available for analytical processing and decision making. The data warehouse defines the manner in which data is:-

- Systematically constructed and cleansed or scrubbed.
- Transformed into a consistent view.
- Distributed wherever it is needed.
- Made easily accessible.
- Manipulated for optimal access by disparate processes.

142.0   Data Warehousing Guide Lines

Although data warehousing is a pervasive term, used throughout the IT industry, there is a lot of misunderstanding as to what a data warehouse actually is.

What is a Data Warehouse

A data warehouse is best defined by the type and manner of data stored in it and the people who use that data. The data warehouse is designed for decision support providing easier access to data and reducing data contention. It isa separated from the day to day OLTP applications that drive the core business. A data warehouse is typically read-only with the data organized according to business rather than by computer processes. The data warehouse classifies information by subjects of interest to business analysts, such as customers, products and accounts. Data in the warehouse is not updated; instead it is inserted (or loaded) then read multiple times.
Warehouse information is historical in nature, spanning transactions that have occurred over the course of many months and years. For this reason, warehouse data is usually summarized or aggregated to make it easier to scan and access. Redundant data can be included in the data warehouse to present the data in logical, easily understood groupings.

The warehouse contains information that has been culled from operational systems, as well as possibly external data (e.g.:- third party point of sale information). Data in the warehouse is stored in a singular manner for the enterprise, even when the operational systems from which the data was obtained store it in many different ways. This fact is important because the analyst using the data warehouse must be able to focus on using the data instead of trying to figure out the data or question it's integrity.

A typical query submitted to a data warehouse is:- "What was the total revenue produced for the X region for product Y during the first quarter?".

To summarize:- A data warehouse is a collection of data that is:-

- Separate from operational systems.

- Accessible and available for queries.

- Subject-oriented by business.

- Integrated and consistently named and defined.

- Associated with defined periods of time.

- Static or non-volatile, such that updates are not made.

143.0 What is a Data Mart?

The term Data Mart is used almost as often as the term data warehouse. But how is a data mart different from a data warehouse??

A data mart is basically a departmental data warehouse which is defined for a single or limited number of subject areas.

Data in data marts need not be represented in the corporate data warehouse, if one even exists. The breadth of data in both data marts and corporate data warehouses should be driven by the needs of the business.

Frequently Asked Questions

How do I differentiate between a data warehouse and a data mart?

A: A data warehouse is for very large databases (VLDBs) and a data mart is for smaller databases. The difference lies in the scope of the things with which they deal.

A data mart is an implementation of a data warehouse with a small and more tightly restricted scope of data and data warehouse functions. A data mart serves a single department or part of an organization. In other words, the scope of a data mart is smaller than the data warehouse. It is a data warehouse for a smaller group of end users.

What are the functional requirements for a data warehouse?

  A: A data warehouse must be able to support various types of information applications. Decision support processing is the principle type of information application in a data warehouse, but the use of a data warehouse is not restricted to a decision support system. It is possible that each information application has its own set of requirements in terms of data, the way that data is modeled, and the way it is used. The data warehouse is where these applications get their "consolidated data."

A data warehouse must consolidate primitive data and it must provide all facilities to derive information from it, as required by the end-users.

Detailed primitive data is of prime importance, but data volumes tend to be big and users usually require information derived from the primitive data.

Data in a data warehouse must be organized such that it can be analyzed or explored from different angles.

Analysis of the historical context (the time dimension) is of prime importance.   Examples of other important contextual dimensions are geography, organization, products, suppliers, customers, and so on.

What are the Data Warehouse Center administration functions?

A: The functions of Visual Warehouse administration are:

1.Creating Data Warehouuse Center security groups.
2.Defining Data Warehouse Center privileges for that group.
3.Registering Data Warehouse Center users.
4.Adding Data Warehouse Center users to security groups.
5.Registering data sources.
6.Registering warehouses (targets).
7.Creating subjects.
8.Registering agents.
9.Registering Data Warehouse Center programs.

Why is there a need to give userid and password to Data Warehouse Center separately when logged on to Windows NT?

A: Because the Data Warehouse Center stores user IDs and passwords for various databases and systems, there is a Data Warehouse   Center security structure that is separate from the database and operating system security. This structure consists of warehouse groups and warehouse users. Users gain privileges and access to Data Warehouse Center objects by belonging to a warehouse group. A warehouse group is a named grouping of warehouse users and privileges that the users authorization to perform functions. Warehouse users and warehouse groups do not need to match the DB users and DB groups that are defined for the warehouse control database.

During initialization, you specify the ODBC name of the warehouse control database, a valid DB2 user ID, and a password. The Data Warehouse Center authorizes this user ID and password to update the warehouse control database. In the Data Warehouse Center this user ID is defined as the default warehouse user.

<u>Improve business with informative data analysis</u>

The Banking Data Warehouse (BDW) Solution from IBM helps people do their jobs more effectively and efficiently. Data is accessible and flexible and presented in a usable format, allowing more time for analysis.

You can access and analyze data such as:

Targeted and direct marketing
Customer retention
Management accounting
Market and credit risk
Fraud detection
Historical business trends

Product gaps and opportunities
Activity and performance
Market segmentation
Competitor products
Actual pricing

Based on the information collected from the data analysis, you can identify
opportunities for:

Focused marketing campaigns
Product customization
Product packaging
Performance tracking
Cross-selling
Promotional pricing
Competitor alliances
Estimation of wallet share

Increase customer loyalty and profitability
Several financial institutions have utilized The Banking Data Warehouse (BDW).

Solution from IBM. For example, a Swedish bank wanted to understand its
customers needs.

BDW helped the bank segment their customers by:

Products used
Behavior
Profitability
Buying preference

Segmenting allowed the bank to identify which options would meet their
customers needs. The solution will help the bank increase customer loyalty and
profitability.

Optimize financial performance
The Banking Data Warehouse (BDW) Solution from IBM can help you
Understand your business and your customers in order to focus on:

Your most profitable customers, optimizing financial performance
Product gaps and opportunities to gain profitability
Key trends to gain a competitive edge
Customer behavior to quickly and effectively promote customized plans that
is the life cycle of a data warehouse?

A: Data warehouses can have many different types of life cycles with independent data marts. The following is an example of a data warehouse life cycle.

In the life cycle of this example, four important steps are involved.

1.Extraction - As a first step, heterogenious data from different online transaction processing systems is extracted. This data becomes the data source for the data warehouse.
2.Cleansing/transformation - The source data is sent into the populating systems where the data is cleansed, integrated, consolidated, secured and stored in the corporate or central data warehouse.
3.Distribution - From the central data warehouse, data is distributed to independent data marts specifically designed for the end user.
4.Analysis - From these data marts, data is sent to the end users who access the data stored in the data mart depending upon their requirement.

<u>What is the Data Warehouse Center control database?</u>

A: When you install the warehouse server, the warehouse control database that you specify during installation is initialized. Initialization is the process in which the Data Warehouse Center creates the control tables that are required to store Data Warehouse Center metadata. If you have more than one warehouse control database, you can use the Data Warehouse Center --> Control   Database Management window to initialize the second warehouse control database. However, only one warehouse control database can be active at a time.

What types of data sources does Data Warehouse Center support?

A: The Data Warehouse Center supports a wide variety of relational and nonrelational data sources. You can populate your Data Warehouse Center warehouse with data from the following databases and files:

1.Any DB2 family database
2.Oracle
3.Sybase
4.Informix
5.Microsoft SQL Server
6.IBM DataJoiner
7.Multiple Virtual Storage (OS/390), Virtual Machine (VM), and local area network (LAN) files
8.IMS and Virtual Storage Access Method (VSAM) (with Data Joiner Classic Connect)

144.0 Questions and Answers

Q. I am using DB2 UDB version 7.2 on Windows NT. In the Data Warehouse Center, I want to create a query the normal way rather than editing SQL ASSIST. I need to put a restriction within a join instead of the where clause and would like to be able to add columns and change the query using the GUI. Is this possible?

A: The GUI is supported by SQL ASSIST. SQL ASSIST does not support the desired action at this time. It appears that the product's inability to meet your request is a limitation with SQL ASSIST. In the version of SQL ASSIST used by Data Warehouse Center, you can only create 'JOIN ... ON' join conditions of the form 'column operator column'.

Frequently Asked Questions

I have installed Data Warehouse Center on AIX and am using Merant ODBC sources to access DB2 and other data sources.
Do I need to make any changes to my .odbc.ini file?

A: Yes, you must change the value of the "Driver=" attribute under the .odbc.ini file. The driver name for AIX is /usr/lpp/db2_07_01/lib/db2_36.o, so the entry should be:

```
[SAMPLE] Driver=/usr/lpp/db2_07_01/lib/db2_36.o
Description=DB2 ODBC Database
Database=SAMPLE
```

For Solaris, this driver name is located at /opt/IBMdb2/V7.1/lib/libdb2_36.so.

Question
Why do I get ESSBASE error 1040008 in the logs, and why does the Data Warehouse Center™ (level U472077) return a DWC7356/rc8410 when I build a cube (load it or calculate it) from Data Warehouse Center?

Answer
To diagnose this problem, you need to consider the errors returned by the components from different products, such as: DB2® Warehouse Manager™ components such as Warehouse Agent, Warehouse Server, Warehouse Client, and End User Interface. DB2 OLAP Server™ components such as the OLAP server, and Application Manager.

Data Warehouse agent returns a DWC7356 error with a secondary return code 8410.

This error is generic and is usually caused because of an issue with an invalid path.

The Warehouse agent reports the error. However, the actual problem could be with the OLAP server component path settings. To diagnose the problem, make a checklist of the installation steps for each product:

Did the application work without problems prior to the occurence of this error, or is it the first time this error occurred?

If so, when did it fail and does it fail only for a new application/database or also for an existing and working OLAP application/database?

If it worked previously, did the failure begin after the DB2 UDB Data Warehouse Center upgrade?

Does the failure occur only with the customer's own OLAP applications, or do the samples fail also?

The Essbase returns a 1040008 error.

When this error occurs, check the path settings. If it is an error reading the message database, check PATH, ARBORPATH, ARBORMSGPATH settings. 1040008 - Essbase returns as "System message - error initializing Essbase." This error message indicates that there was an error while reading the message database. When this happens, do the following: Ensure that the file you are using is not corrupted.

Frequently Asked Questions

Where can I change the isolation level for the Data Warehouse Center Transformer (SQL) and/or Source Database, and at what level will this happen?

A: The Data Warehouse Center agent reads with the cursor stability isolation level by default. This cannot be changed, but you might consider the following as a possible workaround:

Define a user defined procedure (UDP) for the import step and cascade on success to the SQL step. Then both steps are managed in the same process and the SQL step will not be started until the import is successful. If you do not want the SQL step to start until the import is finished, then cascade on success is one way to do it.

If the import step must run outside of the warehouse process, then you should use another mechanism to start the SQL step. The file wait program could be used. After the Import completes successfully, create a file. Then use the file wait program to poll for the file and cascade to the SQL step. Do not forget to erase the file waited upon after the SQL step or before the import.

Reference:78022 / 98908   Date: 17 Dec 2002   [Return to questions | Return to main page]

How do I set the log level higher for more detailed information within Data Warehouse Center 7.2?

A: Within DWC, log level capability can be set from 0 to 4. There is a log level 5, yet it cannot be turned on using the GUI, but must be turned on manually. A command line trace can be used for any trace level, and this is the only way to turn on a level 5 trace:

1.Go to start, programs, IBM DB2, command line processor.

2.Connect to the control database:

  db2 => connect to Control_Database_name

3.Update the configuration table:

  db2 => update iwh.configuration set value_int = 5 where name = 'TRACELVL' and (component = '<component name>')

  Valid components are:

    a.Logger trace = log
    b.Agent trace = agent
    c.Server trace = RTK
    d.DDD = DDD
    e.ODBC = VWOdbc

4.For multiple traces the format is:

  db2 => update iwh.configuration set value_int = 5 where name = 'TRACELVL' and (component = '<component name>' or component = '<component name>')

5.Reset the connection:

  db2 => connect reset

6.Stop and restart the Warehouse server and logger.

7.Perform the failing operation.

8.Be sure to reset the trace level to 0 using the command line when you are done:

  db2 => update iwh.configuration set value_

145.0 Install and Configure ESE (EEE) revisited

1. **Make the file system for the DB2 instance's home directory available across NFS to all SP Nodes.**
   Smitty nfs
   → Network File System
   → Add a Directory to Exports List
   Specify:- PATHNAME of directory to export ………… /home/instance owner
           HOSTS allowed root access   ………… hostname2/hostname3/ ……….

2. **Installing UDB EEE**
   Log in as root on the first node xxxxxxxx
   Change Directory to where the DB2 images are locarted.
   Run Smitty installp, then specify . (dot) as the input directory.
   Chose License for DB2 UDB EEE
   Run the installp instruction
   Installp –acqNQqwx –d. 'db21_08_00.xsrv 8.0.0.1'
   Or run DB2setup off the CD

3. **Add a Group and User**
   Smitty group
   → Add Group
   Group Name   db2asgrp
   Add User
   Smitty User
   User Name   db2inst1
   Primary Group   db2asgrp
   Home Direcotry   /home/xxxxxxxx/db2inst1

4. **Add entries to the /etc/service file.** Entries must be made to reserve the ports and specify the service names.
   If you are using 4 database partitions you per SP Node need to reserve 4 ports per SP norde.
   To support HACMP you need to reserve another 4 ports.
   Eg.
   /ctc/scrviccs
   DB2_db2inst1   50000/tcp
   DB2_db2nst1_END   50007/tcp
   This means ports 50000 to 50007 are reserved for DB2.

5. **Create the Instance.**
   Logged In as root, create the instance:-
   xxxxxx> /usr/lpp/db2_08__00/instance> ./db2icrt –udb2inst1 db2inst1
   Completed Successfully message should display.

6. **Add the profile entries**
   Entries need to be added to t he .profile of db2inst1 so that environment variables and DB2 profile variables are set.
   Add these lines to:- /home/xxxxxx/db2inst1/.profile:
   .~/sqllib/db2profile
   #Set the default DB
   db2set –1 db2inst1 db2dbft=tcpd30 ← database name.

7. **Edit the db2nodes.cfg file**
   The file db2nodes.cfg in the directory $INSTHOME/sqllib defines which database

partition servers will be started when a db2start is issued.

1 xxxxxxx 0   ssssssss
2 xxxxxxx 1   ssssssss

Col 1 = DP (Database Partition). Col 2 = Host Name, Col 3 = Logical Port, Col 4 = Switch Name

8. Creating .rhosts entries

Before issuing a db2start, we have to make sure that remote execution permission is defined across the SP nodes. In other words, if we execute rsh xxxxxxx date from xxxxxxx01 we will

Not be prompted for a password. This needs to be for all the network interfaces defined in db2nodes.cfg.

This permission is set by adding these lines to .rhosts in /home/xxxxxx01/dnb2inst1 on xxxxxxx01:-

xxxxxxx01 db2inst1

ssssssss01 db2inst1

xxxxxxx05 db2inst1

ssssssss05 db2inst1

etc etc.

The permission on this file must be:- chmod 600 .rhosts

Because /home/xxxxxx01 is a file system that is shared across the SP nodes, we only have to create these entries once.

9. Setting up the db2diag.log

This file is by default stored in the $INSTHOME/sqllib/db2dump directory, which is NFS shared to all the SP Nodes. It is a good idea to change this path so that the db2daig.log is written locally on each SP node. To do this, logged in as db2inst1 at xxxxxx01:-   db2 update dbm cfg using diagpath /tmp/db2   Where /tmp/db2 is a file system that exists locally on each SP node (not NFS shared). Since this Parameter (diagpath) is defined at the database manager level, this command is executed once for the DB2 instance. Note:- All the DP (database partitions) on the same SP node will write to the same file, so output From DP 1,2,,and 4 will all go to:- /tmp/db2/db2diag.log on xxxxxx01.

10. Setting up the syslog.conf

Log on as root. (this is for information messages independent of the DB2 Instance. These messages are captured by configuring the AIX syslog daemon. Add this line to /etc/syslog having logged ion as root:-   User.warn /tmp/db2/syslog.db2   Create to /tmp/db2/syslog.db2 file   Touch tmp/db2/syslog.db2   Fine the process id (pid) of the syslogd process:-   Ps –ef | grep syslogd   Send a –1 signal to this process using kill:-   Kill -1 <syslogd-pid>

Run this set of commands at each one of t he SP nodes.

11. Starting the instance

To do this, logged in as db2inst1 enter:- db2start.

You should get numerous messages saying DB2stgart processing was successful.

12. Creating the database

db2terminate
export DB2NODE=1

db2 –v "create db tcpd30 on /DB_LOG catalog tablespace managed by database using (device '/dev/rlv_n01_01_101' 10240)"

I have explicitly set DB2NODE so we will know which database partition the System Catalogs will be stored on.
The database is created on /DB_LOG so that the file systems for the logfiles will be used eg:-
/DB_LOG/db2inst1/NODE0001 for the first DP.

13. Set archival Logging

To enable archival logging we set the logretian switch on at this point, this will force a backup, which will be much faster now rather than after we load data. To run db2 update db cfg at all DPs we issued:-

Db2 all "|| update db cfg for tcpd30 using logretain on"

The double pipe means the command will be run in parallel on all DPs.

14. Backing up the database:-

db2_all "<<+1<db2 backup db tcpd30 to /backup"
db2_all "<<-1<||db2 backup db tcp3d to /backdb"

+1 means run the command at DP 1
-1 means run the following in parallel at all DPs except DP1

15. Creating the Node Groups:-

create nodegroup xx_large on nodes (2 to 16);
create nodegroup xx_small on nodes (1);

16. Create the Tablespaces

1. create tablespace ts_1 in xx_small managed by database sinf (device 'dev/rlv/01_01_101' 10240);
2. create tablespace ts_2 in xx_large managed by database using (device 'dev/rlv01_01_210' 15360.
3. etc etc.

146.0   INSTALLING DB2 for HP-UX

Use db2setup and HP-UX's native install program

1. Before installing your DB2 for HP-UX check to see if you need to update your system's kernel configuration parameters. You must then reboot your machine after updating any kernel configuration parameters.
2. The msgsem parameter must be set no higher than 32767.
3. The msgmb and msgmax parameters must be set to at least 65535.
4. The shmax parameter should be set to 134MB or 90% of t he physical memory in btes.
5. Change parameters in seqence.

To change a Kernel parameter perform the following:-

Step 1:- Enter SAM to start the System Administration Manager (SAM) program.
Step2:- Double click the kernel configuration icon.
Step3:- Double click the Configuralbe Parameters icon.
Step4:- Double click on the parameter you want to change and enter the new value in the Formula/Value field.
Step 5:- Click on OK.
Step 6:- Repeat these steps for all the kernel configuration parameters you want to change.
Step7:- When you are finished setting all the kernel configuration parameters select:-
        Action→ Process New Kernel from the action menu bar.
The HP-UX operating system automatically reboots after you change the values for the kernel configuration parameters.

Installation

To install DB2 for HP-UX using db2setup:-

1. Log in as a User with root authority.
2. Insert and mount your DB2 product CD-ROM.
3. Change to the directory where the CD-ROM is mounted by entering the CD/ROM. Command where cdrom is the mount point of your product CD-ROM.
4. Enter the ./db2setup command. The DB2 Setup. The DB2 Setup Utility window opens.
5. Select Install and press enter. Install DB2 V8 window opens
6. Select the products you want and are licensed to install. Press TAB to move between available options and fields.
7. You can choose Customize to view and select components that will be installed.
8. Select OK to continue with the installation process or Cancel to go back 1 window
9. Select Help for more information or assistance.
10. When installation is complete your DB2 software will be installed in the:- /opt/ IBMdb2/V8.1/ directory.
11. If you did not install DB2 Tools you can verify installagtion by creating and connecting to the Sample database.

Installing Clients

After updating your kernel configuration parameters and rebooting your system you can install your DB2 Client.

Log in as user with root authority.
Insert the appropriate CD ROM.

Change to the directory where the CD-ROM is mounted by entering the cd/cdrom command where /cdrom is the CD ROM mount point.

Change to /cdrom/db2/hpux11 for HP-UXC

Enter the ./db2setup command, The install window opens.

Select the product you want to install.

Press TAB to move between options and fields.

Select Customize to vie w and change components.

Select OK to continue or Cancel to go back.

When the installation is complete, the DB2 software installed in the DB2DIR
 Directory.
Where DB2DIR = /usr/lpp/db2_08_01 on AIX

= /opt/IBMdb2/V81. on HP=UX

You can add additional components after your initial install by entering for HP-UX/opt/ IBMdb2/V8.1/install/db2setup

After install, test accesss to a remote server

147.0 Communication DB2 UDB EEE on Windows NT

A: DB2 UDB EEE runs on a cluster of Windows NT systems interconnected by TCP/IP or VIA. For small systems, like two to four nodes of uniprocessors or two nodes of four-way SMP, you can use a fast Ethernet connection. For a larger number of clusters, there are several vendors who provide higher speed interconnects.

148.0   DATA SHARING and Parallel Sysplex revisited

Data Sharing allows applications running on multiple DB2 Subsystems to concurrently read and write to the same data sets. Simply stated, Data Sharing allows multiple DB2 Subsystems to behave as one.

Data Sharing is optional it need not be implemented.

The primary benefits of data sharing is to provide increased availability of data.

What are Sysplex and Parallel Sysplex??

A Sysplex is a set of OS/390 images that are connected by sharing one or more Sysplex Timers. A parallel Sysplex is a basic Sysplex that additionally shares a coupling facility whose responsibility is to provide external shared memory and a set of hardware protocols that allow enabled applications and subsystems to share data with integrity by using external shared memory. Parallel Sysplex enhances availability by providing customers with the ability to non-disruptively one or more OS/390 images and/or CECs from a configuration to accommodate hardware and software maintenance.

Data sharing consists of a complex combination of hardware and software. To share data, DB2 subsystems must belong to a pre-defined data sharing group. Each DB2 subsystem contained on the data sharing group is a member of that group. All members of a data sharing group access a common DB2 Catalog and Directory.

Each data sharing group is a OS/390 Cross-system Coupling Facility (XCF) group. XCF The group services provided by XCF enable DB2 data sharing groups to be defined. In addition XCF enables the data haring environment to track all members contained in the Data sharing group. A site may have multiple OS/390 Sysplexes each consisting of one or more OS/390 Systems. Each individual Sysplex can consist of multiple data sharing groups.

DB2 requires a Sysplex environment that consists of the following:-

1. One or more Central Processor Complexes (CPCs) that can attach to a coupling facility.

2. At least one coupling facility. The coupling facility is a component that manages the shared resources of the connected CPCs. DB2 uses the coupling facility to provide data sharing groups with coordinated locking, bufferpools and communication. MVS V5 and up is required to install a DB2 coupling facility.

3. Application Impact - none. Attachment Interface impact – none.

4. Sysplex distributed access using both public and private protocols can be made to a data sharing group.

5. Application Support Impact – substantial.

The DB2 Coupling Facility

DB2 uses the coupling facility to provide inter-member communications.
The primary function of the coupling facility is to provide is to ensure data availability while maintaining integrity across systems.

To do so, the CF requires three structures to synchronize the activities of the data sharing group members.

    a. Lock structures – control global locking across the data sharing group.
    b. List structure – enables communication across the Sysplex environment.
    c. Cache structures – provide common buffering for the systems in the Sysplex.

Defining the Coupling Facility

1. A coupling facility is defined using Coupling Facility Resource Management (CFRM). CFRM is created by the IXMIAPU utility. The CFRM is used to identify:-

2. Each individual coupling facility.

3. Each structure within the individual couple facilities.

4. Space allocated to these structures.

5. Ordered preferences and which coupling facility is used to store this ordered preference structure.

6. Under exclusion list, which defines the structures to keep separate from this structure.

Consider the following:-

Global Lock Management
Global inter system communication (sca)
Sharing buferpools. Group bufferpool duplexing
Data sharing naming conventions
Data sharing administration
Data sharing group creation.
Backup and recovery.
Subsystem availability.
Monitoring data sharing groups
Coupling facility recovery

149.0   The DB2 UDB Version 8.1 enhancements

1. Online load enables load at the table level allowing access to other tables in a multi table tablespace. Load Query – returns status of target table. Load Wizard available.
    Load Wizard includes:- Memory Visualizer
    Redistribute wizard
    Backup and restore wizard
    Configure database logging wizard
    Add partition wizard
    Alter database Partition Group
    Storage Management View
    Design Advisor (enhanced DB2advis).

2. Has Self Management and Resource Tuning (SMART). Manage more data with less effort.

3. Flush Package Cache. A new FLUSH PACKAGE CACHE lets you remove Cached dynamic SQL statements by invalidating them.

4. Logging Enhancements.
    DB2 V8.1supports dual logging. You can now specify the MIRRORLOGPATH DB CFG parameter. This parameter replaces the DB2NEWLOGOATH2 parameter. The maximum amount of log space has been increased to 256GB from 32GB. Infinite logging a current unit of work to span primary and archive logs. The BLK_ON_LOG_ DISK_FUL registry variable has been replaced with a new database confiig parameter BLK_LOG_DSK_FUL, which tells DB2 not t o fail on a log full condition, instead, DB2 will retry the write to the log every five minutes giving you time to resolve the disk full condition.

5. Backup and recovery enhancements. A new table space history file identifies log files needed for a particular table space recovery. Log files that are not needed are skipped, resulting in faster table space recovery. You no longer have to specify Coordinated Universal Time for rolling forward to a point in time. Instead, you cab now specify local time, eliminating confusion converting local time to CUT time.

6. DB2DIAG.log has been split into two files. 1. The Admin Notification Log and 2. DB2DIAG.log for errors. 1. Is for use by DBAs and Sys Admin. The level of info is controlled by a new NOTIFYLEVEL database parameter. Detailed diagnostic info is still written to the DB2DIAG.log.

7. There is a new database maintenance mode. This new feature is like the ACCESS(MAINT) command available on z/OS and OS/390. You can now use DB2

V8.1 QUIESCE command to force all users off an instance or database and place these objects in quiesce mode so maintenance can be performed.

8. REORGCHK enhancements now includes an ON SCHEMA option to specify a particular schema.

9. RUNSTATS enhancements can now collect statistics on a table and on index and table relationships. RUNSTATS can now also accept lists of indexes and columns on which statistics are to be collected.

10. Management by exception monitoring – The new Health Monitor is a server side tool that can raise alerts when predefined thresholds are exceeded or when it detects an instance is down. Based on the alert generated, the Health Monitor can send an e-mail or pager address, trigger a script, or run a task through the new Task Center.

11. Event Monitor and Snapshot Monitor – You can create Event Monitors to write to tables instead of to files or pipes. You can now write customized queries to these tables containing Event Monitor data. And you can now take snapshots through SQL Table functions. This output can be joined to other tables, previously that output could only be embedded in programs thsat used th Administrative API.

12. Application Development – DB2 V8.1 takes existing XML support to new levels. The XML Extender V8.1 supports Web services with the Web Services Object Runtime Framework WORF Tool Set. DB2 V8.1 incorporates WebSphere Studio and Web Sphere Application server, to create publish and maintain dynamic Web applications.

13. SQL –The new INSTEAD OF Trigger extends the ability to update views.

14. Multidimensional Clustering is a new clustering technology. Provides for automatic continuous clustering of data along multiple dimensions. And MDC does not require database maintenance operations such as reorgs. Cells are made up of unique combinations of dimension values composed of Blocks of pages where a block is a set of consecutive pages on disk. Slices are sets of blocks containing pages having certain key values of one of the Dimension block indexes Dimension Block Indexes are indexes that are automatically created for each dimension specified. Composite block indexes are automatically created indexes that contain all dimension key columns. They are used to maintain clustering over of data over Insert and update activity.

15. MQT Materialized Query Tables – new for V8,1 – use summarized data to provide high performance joins. Automatic summary table (ASTs). AST has a FULL SELECT clause that contains GROUP BY clause summarizing data from tables referenced in thew FULLSELECT

16. VALUE COMPRESSION and COMPRESS SYSTEM DEFAULT of the CREATE TABLE statement. Reduces storage requirements.

17. Catalog and authorization caching – for databases with multiple partitions an extension of the catalog cache will be provided at each partition.

18. Asynchronous I/O enhancements:- V8.1 exploits AIX page cleaning Performance.

19. Java stored proceduere performanc implemented usinf single thread model.

20. Connection Concentrator many connections can share the same coordinating Agent,.

21. Online table reorgs.

   Type 2 indexes eliminates next-key share locks

22. Online index reorgs.

23. Set configuration parameters online.

24. Online buffer pool enhancements – create, drop or alter buffer pools online

25. INSPECT lets you check the architectural integrity of the table spaces and tables online. It can also be used to identify types of indexes on a table.
    Also DB2 Trace facility can capture important diagnostic info. for DB2 support.

    DB2SUPPORT provides DB2 with a "support bundle" of diagnostic info.

26. DB2dart – database Analysis and Reporting tool is now officially supported
    In V8.1 used also to diagnose structural problems with underling DB2 objects and Repair damaged objects.

27. In a nut shell:- easier to manage, makes DBAs more productive. New Wizards take the pain out of work. Application development tightly integrated.

<u>150.0   CATALOG THE NODE</u>

```
CATALOG--+---+--TCPIP NODE--nodename---REMOTE---hostname---->

>-----+-----SERVER---service_name--+---+--------------+---->

>-----+--------------------+---+------ ---------------+---->

'REMOTE_INSTANCE--instancename--'   'SYSTEM--system-name'

>-----+-----------------------+----------------------->
      'OSTYPE--operating-system-type--'



 >-----+--------------------+----------------------------->
      'WITH--"comment-string"--'
```

For example:

CATALOG TCPIP NODE evoke REMOTE machine_name SERVER port_
number REMOTE_INSTANCE instance_name

CATALOG THE DB:


```
>>-CATALOG----+-DATABASE-+--database-name----+--------+---->
             '-DB-------'                    '-AS--alias--'

>-----+----------------+------------------------------->
      +-ON--+-path--+------+
      |     '-drive-'      |
      '-AT NODE--nodename--'

>-----+--------------------------------------------+>
      |                 .-SERVER---------------------------
-------.  |
      '-AUTHENTICATION--+-CLIENT----------------------
-------+--'
                        +-DCS------------------------------
-------+
                        +-SERVER_ENCRYPT-------------------
-------+
                        +-DCS_ENCRYPT----------------------
```

```
------+
                                +-KERBEROS TARGET PRINCIPAL--
principalname--+
                                '-DCE SERVER PRINCIPAL--
principalname------'

>-----+----------------------------+-------------------------><
      '-WITH--"comment-string"--'
```

For example:

CATALOG DB EVKREBDB AS EVKREBDB AT NODE evoke

CATALOG DCS ENTRY:

```
>>-CATALOG DCS----+-DATABASE-+--database-name--------------->
                  '-DB-------'

>-----+-----------------------+---+-------------------+----->
      '-AS--target-database-name--'   '-AR--library-name--'

>-----+---------------------------+------------------------->
      '-PARMS--"parameter-string"--'

>-----+---------------------+-----------------------------><
      '-WITH--"comment-string"--'
```

For example:

CATALOG DCS DB EVKREBDB as EVKREBDBDCS

151.0   What is Performance Tuning?

In a nutshell, Performance Tuning is a change or changes to an entity that lowers that entity's execution time.

However, there are several areas where changes may be applied in order to improve execution time, and broadly speaking these are:-

1. OS/390 MVS Operating System.
2. DB2 Sub-system.
3. The Network Environment.
4. Database Design.
5. Application Program Design.

Due to the unique nature of the Operations at some companies, I will concentrate only on points 4 and 5. Point 4 (Database Design) will take up approximately 25% of the tuning effort and Point 5 (Application Program Design) will take up approximately 60% of the tuning effort. These two functions alone will then take up 85% of the tuning effort!!!

How do you Performance Tune at a High Level?

You performance tune at a High Level by:-

1. Devising a Performance Tuning Plan and getting the Plan approved by all the parties concerned i.e. Management, DBA Group and Application Developers.
    All parties have to agree on all points of the plan otherwise the effort leaves open the risk of failure.

2. By listing in detail each step required for each Application Program and / or Query (entities). Start with the worst offenders as candidates to be tuned first.

3. By keeping track of every change made to each Application Program and/or Query (entities).

4. By running each entity **before** the changes are applied and logging the performance results in an Excel Spreadsheet or comparable Matrix.

5. By running each entity **after** the changes are applied and logging the performance results in an Excel Spreadsheet or comparable Matrix. This Matrix could be the same one referred to in point 4, as long as the Date and Time of each test is logged.

6. When ever and entity is run before and after changes are applied, these runs must take place in the same environment, preferably at the same time of day.

7. The watchword for this type of Performance Tuning is "Systematic".

How do you Performance Tune at a detailed level?

1. Select an Application Program for tuning. Run REORG and RUNSTATS on the Appropriate objects (Tables and Indexes).
2. Analyze each Access Path chosen by the DB2 Optimizer for each SQL Statement. Set up a full sized PLAN TABLE.
3. Run DB2 EXPLAIN to accomplish Point 2, for each SQL Statement. Log the Results.
4. If in your analysis, you see that certain I/O is not necessary, make a note of this.
5. Run the job that executes the relevant Application Program/Query.
6. Log the results (Elapsed Time and CPU time).
7. Apply all the changes you think are necessary to the SQL, based on your previous analysis,
8. Run REORG and RUNSTATS as in Point 1.
9. Run the job that executes the relevant Application Program/Query.
10. Log the results (Elapsed Time and CPU Time).
11. If there is an improvement by more than 20% then move on to the next entity, otherwise revise your changes, seek advise, try again. At this point it may be beneficial to execute the changes one at a time, logging each of the results.
12. Make sure you eliminate the unnecessary I/O identified in Point 4.

What do you look for when Performance Tuning?

1. Are the number of SQL Statements the minimum required. More often than not you will find unnecessary SQL Statements when you go through the code.
2. Based on the results of the EXPLAIN, is the Program making maximum use of Indexes, or is it using Table space scans.
3. Are the number of SELECTS the minimum needed for a row? You will invariably find unnecessary SELECTS in a program. This will have a huge negative effect.

    Example:- a. Average 10 transactions a second.
    b. Average 3 Selects for same row (2 unnecessary).
    c. 15 hour average processing day.
    d. $10 * 2 * 60 * 60 * 15 = 10,800,000$ unnecessary calls to DB2.

Number of rows retrieved:- The actual number of rows that need to be updated, inserted or deleted in an application is based on a combination of design and business requirements, however the actual number of rows retrieved is more a programming than a SQL issue.

The overhead for retrieving one unnecessary row can be very high, depending on how many times it is done. Every unneeded row that is brought into a program generally causes the following unnessary events to occur:-
    a. GETPAGEs which might require additional I/O for both data and index.
    b. Movement of columns selected from page buffer to work area.

    c. Transferring data cross memory to the program.

    d. Extra code execution in the program to test whether or not to filter.

One simple technique to find out whether unneeded row retrieval is occurring is to verify code by looking for IF statements after an SQL statement that do filtering that should have been done within the SQL.

4. Number of columns retrieved:- Every column retrieved has to be moved from the buffer pool to a user work area. Even though there are times when individual columns are passed together in a single cross-memory call, it is best to view this as one cross-memory move per column, as this is generally the case. There is also overhead when all the columns "retrieved" get to the stage 2 relational data services layer in DB2, only to be filtered out. In many cases that could have been done in the stage 1 data manager layer.

Statistics show that industry wide, only 0.01% of application programs do not unnecessarily retrieve more columns that needed. So that means over 98% of all Application programs are retrieving columns they do not need. A huge overhead!! Also, columns not needed are moved to host variables that never get used, and 99% of Application Developers do not go back and eliminate these column retrievals because they say they are under pressure. Some do not even consider it an issue!!!

5. Unnecessary Repetitive Processes:- Many times a critical process is repeated for every occurrence of an event. When a condition is met, the loop should be terminated. But this is not generally the case. Processing continues to the end of the data. This means that a huge overhead could be encountered when ever a Perform Loop is executed.

6. Referential Integrity:- Implementation of referential integrity (RI) may have implications from a CPU resource perspective, especially when RI is used across the board without Performance Certification, that is a design default method. This applies to both defined (declarative) and application program RI, where all relationship checking is done in the program code. In general, RI is not efficient at all for a high performance application. You need to be precise in the design stage for the use of RI. Also the data types used in RI are very important. For example a CHAR type is nowhere near efficient as a SMALL INT.

7. Sorting:- Many sorts are not needed, yet never get removed or corrected. Many times an index not only can do away with the sort but can dramatically improve response time and reduce overhead.

8. Physical Design Issues:- Unnecessary CPU overhead can sometimes be the result of poor index design or lack of appropriate indexes.

9. Application Design Issues:- Application design can obviously contribute to unnecessary CPU overhead, specifically when it comes to SQL syntax and guidelines.

Stories about the application program that used to run in two hours and now runs in one minute are common place.

10. I/O:- The biggest "enemy" to performance is I/O. Therefore a very careful look at any and all I/O must be conducted. I/O takes the form of SELECTs, INSERTs, UPDATEs, DELETEs and non SQL related I/O. So "if it moves analyze it" must be your watch word.

11. Thoroughly analyze the results of a query. Sometimes, for reasons best known to itself, DB2 will choose an alternative access path, down grading performance. In these cases first try a re-bind, before delving further into your analysis.

12. Minimize the number of rows in a Join:- Joining too many tables in one query will affect performance. Officially IBM states that the maximum number of tables that can be joined in a single SQL statement is 225, however the practical limit is under 10. NEVER use a Join statement without a predicate!!!

13. Joins:- The are several "DOs" and "DON'T" regarding Joins and this is currently beyond the scope of this exercise. I will return to Joins at a later date, meanwhile consult the relevant IBM Manual.

14. Always use Index Access except for very small talbes.

15. Use Clustered Indexes where ever possible.

16. Use the FREEPAGE and PCTFREE parameters to reduce the frequency of REORGs (FREEPAGE) and increase the space available for inserts (PCTFREE).

17. Systematic Rebinding:- A good practice is to rebind plans after executing REORG and RUNSTATS,

18. Lower the Filter Factor by using redundant predicates. When doing this be sure to document the reasons why you used the redundant predicates.

19. Place the most restrictive predicates at the beginning of your WHERE Clause list.

20. Rebalance the partitioning keys. Check if the data is skewed.

21. Determine Foreign Keys without an index. All Foreign Keys should have indexes.

22. Lock Escalation:- Lock Escalation can be turned off by setting the LOCKMAX parameter to zero. A simple ALTER statement can effect this change. The object must be stopped and started before and after the ALTER.

One Liners for Tuning the Database Design

If Necessary:-

Modify the Logical Model

Modify the Physical Model

Modify and issue new DDL

Use partitions for large tables

Spread non-partitioned objects over multiple devices using PIECESIZE

Add Indexes

Change Indexes

REORG Table and Index spaces

REBUILD Indexes

De-normalize

Add redundant tables

Record all changes

One liners for tuning Programs

If Necessary:-

Perform SQL Tuning

Use triggers to enforce rules

Implement stored procedures to reduce network traffic

Tune COBOL Programs

Use RUNSTATS

Execute EXPLAIN and REBIND
Change locking strategies

Change Catalog statistics and rebind

Implement Optimizer influences

Use testing tools

Record all changes

Test and Production

To really reflect the Production environment, the pre-production test environment needs to duplicate the production environment as closely as possible to get accurate statistics. Consequently, it benefits all concerned to make sure these two environments reflect each other. Don't forget many ZPARMS have to be tuned in addition to all the "other" Pools and Parameters.

Summary

When you look for improvements you should think about each little pressure point in DB2. Ask Questions. Is this the best? Can it be better? Is it hurting? If you change it, can you allow more work?

Look for simple things that are wrong, they usually case the most problems.

Look for excessive use of Stored Procedures!!!! Are they justifed. Check the Strored Procedure parameters,, such as STAY RESIDENT, PROGRAM TYPE SUB etc.

Recommendations

1. Study this document.

2. Modify and finalize as necessary.

3. All parties must agree on the final approach.

4. Allocate resources.

5. Start implementation of the strategy as agreed.

6. Measure to results.

7. Show the results to all interested parties.

8. Implement the changes to production.

9. Measure the results. Compare before and after results.

===

152.0 Disaster Recovery

Use logs as a primary means of ensuring the consistency and recoverability of data. In these notes I will explain how logs are managed and how DB2 UDB can use a user managed exit program to archive and retrieve database log files. Also, I will provide a step by step example to explain how to modify, compile and test the db2uext2.cdisk sample user exit program that comes with the installed DB2 UDB server product. These notes are not the solution for XYZ Inc., but merely a guide which can be used for useful tips and information.

How does a user exit work with DB2 UDB?

The idea behind using a user exit program with DB2 UDB is to provide a method of archiving and retrieving database log files to enable log redundancy and movement away from unstable or volatile media. It is also important to know that you may implement other operations besides archiving and retrieving logs with the user exit, based on you specific needs.

Having a user exit strategy with DB2 UDB nay not recover 100% of your transactions if a database needs to be recovered using archived log files. A user exit program is merely a means to provide more protection for your existing log files by copying the log files to a safe location. It is one part of a data integrity strategy, but it is an important part.

After compiling a user exit program, the db2uext2 executable is placed in a directory where the database manager can find it. The directory is, <instance home> sqllib/adm. The database manager will not call db2uext2 unless it is aware that a user exit program is available. The only way the database manager will know that the db2uext2 program can be called is by setting the database configuration parameter USEREXIT to ON. Once this parameter is set and the instance is recycled the database manager will make a call to the user exit program every five minutes, to check for log files that can be archived to the program specified archive directory.

If a database recovery is necessary, the database manager will call db2uext2 during the roll forward operation to copy the archived log files back to the active log directory. He log files are then reapplied to the restored database.

Let us take a look at the calling format made by the database manager to the user exit program:-

Db2uext2 –OS<os> -RL <release> -RQ<request> -DB<database> -NN<nodenumber> - LP<log path> -LN<log name> {-AP<password>

Where:-
Os = Operating System, release = DB2 release, request = 'ARCHIVE or 'RETRIEVE'
DbName = database name, nodenumber = node number, log path = log file path, logname =

log file name, password = ADSM Password (Optional).

Log files are archived and retrieved from disk with the following naming convention:-

Archive:- archive path + database name + node number + log file name
Retrieve:- retrieve path + database name + node number + log file name

Example:- archived to:- logs/sample/node0000/S0000001.LOG
               Retrieved form:- logs/sample/node0000/S0000001/LOG

Here's how the logic flows within the user exit:-

     a.   Install signal handlers
     b.   Verify number of parameters passed
     c.   Verify action requested
     d.   Start Audit Trail if requested.
     e.   Take one of the following pats based on the action requested
            1.   If the requested action is archive a file, copy the log path to the archive path
            2.   If the log file is not fond – log the error and exit
            3.   If the requested action is retrieve a file, copy the log file from the retrieve path to the log path.
            4.   If the file is not found – log the error and exit.

     f.   End the Audit trail if requested.
     g.   Exit with the appropriate return code.

It is possible for you to manually call the user exit program to archive a lg file, but it is better to use the ARCHIVE LOG command so that mistakes are not made when specifying the parameters above.

Log file terminology

The basic function of a user exit program with DB2 is to copy log files to and from the active log directory. I mention some terminology here to clarify where the actve log directory is located and the states of the database log files.

Active log directory

The Active log directory is located in your database directory.

/DB2INST1/NODE0000/SQL000001/SQLOGDIR

Within the above several logs may exist:-   S0000000.LOG, S0000001.LOG, S0000002.

LOG.

Log file states

Within the active log directory log files can either be active logs or online archived logs. Active logs are those logs needed by DB2 for current transaction processing and crash recovery. Online archived logs are no longer needed by DB2 UDB for reqular processing but may be needed during a database recovery. When implementing a user exit program, these online archived logs should eventually show up as a copy in your archived log directory.

Since the purpose pf a user exit program with DB2 UDB is to copy database logs to an archive directory, you will end up with duplicate log files in your active log directory, by default. SQLOGDIR. You may consider removing the duplicate online archived logs to free up the file system space. Be very careful to verify that these logs were successfully copied to the archive directory before removing them from the database directory. You must also make sure they are no longer needed by the database manager for crash recovery. To determine which log files are no longer needed for processing in your active log directory, check the database configuration as follows:-

Db2 "get cfg for samledb"

The database configuration output from this command will include the first active log file, for example:-

First active log file = S000000.LOG

The log file S000009.LOG shown above is the currently active log for the database. Any log files that are numbered less than this log file are considered online archived logs.

Here is an example:-

In this scenario the active log directory holds log files S000000.LOG to S000009.LOG and the archive log directory holds S000000.LOG to S000008.LOG. Because S000009.LOG is the first active log file, S000001.LOG to S000008.LOG can be deleted from the active log directory to free up disk space. The S000009.LOG file must be left in the active log directory since it is still being used for current transactions.

The database history file can also be checked to see which log files are no longer required in the active log directory. The following command willl list database backup information:-

Db2 "list history backup al for database sample"

Here is the output from the above command:

List History File for sample
Number of matching file entries = 4
<u>Op</u> <u>Obj</u> Timestamp+Sequence <u>Type</u> <u>Dev</u> <u>Earliest Log</u> <u>Current Log</u>

B   20040416162026001   F   D   S0000010.LOGS0000014.LOG

Contains 2 table spaces
00001 SYSCATSPACE
00002 USERSPACE1

In the above output, the earliest log above will signify that any log after and including the S0000010.LOG is needed. Any log before S0000010.LOG can be safely deleted. Again, it is important to verify that a copy of the log files exist in the archive log directory before deleting those logs from the active log directory.

Although manually deleting log files from the active log directory is possible, a safer way to remove online archived log files from the active log files is through the 'prune log file' Command. This command can be used to delete log files S000000.LOG to S0000008. LOG:-

Db2 "prune logfile prior to S000009.LOG"

Note:- Depending on the recovery strategy, there may be scenarios where previous roll forward operations are performed on the database. Old log files in the archive directory may be overwritten with newer log files with the same name, thereby preventing point in time recovery of a database using old log files. It is important for the programmer or DBA writing the user exit to take such situations into account.

Setting up a user exit - example

1. Create the following directory structure:-

    /mylogs/SAMPLE/NOD0000

    Then give recursive permissions to each directory in the structure:-

    chmod –R 777   /mylogs

2. Copy the db2uext2.cdisk to a working directory and give file permissions:

    cp /home/db2v8_32/sqllib/samples/c/db2uecxt2.cdisk
/home/db2v8_32/db2uext2.c

chmod 777 /home/db2v8_32/db2uext2.c

3. For this example the following part of the user exit program should be updated to Reflect the path /mylogs/

    ```
    #define ARCHIVE_PATH "mylogs/"
            /* path must end with a slash */
    #define RETRIEVE_PATH "/mylogs/"
            * path must end with a slash*/
    #define AUDIT_ACTIVE 1
            /* enable audit trail logging */
    #define ERROR_ACTIVE 1
            /* enable error trail logging */
    #define AUDIT_ERROR_PATH "mylogs/"
            /* path must end with a slash */
    #define AUDIT ERRORATTR "a"
            /* APPEND TO TEXT FILE */
    #define BUFFER SZE 32
            /* # of 4K pages for output buffer */
    ```

4. Make sure you have a C compiler installed on the system and that your environment has the compiler's library and path in it.

5. From the command line build the db2uext2.c program:-

    ```
    Cc –0 db2uext2  db2uext2.c
    ```

    Once the program is compiled, a db2uext2 executable will be created.

6. Move the db2uext2 executable to the sqllib/adm directory.

    Prepare the database for user exit.

7. Create the SAMPLE database (if not already created) with the db2sampl command. This will allow you to use the sample tables for the examples below.

    ```
    Db2sampl
    ```

8. Update the database configuration file so that the user exit facility is turned on for the database:

    ```
    db2 "update db cfg for sample sing userexit on"
    ```

db2stop force db2start

Note that the logretain switch does not have to be set to enable roll forward recovery if you are using the user exit facility.

9. For testing purposes, you can encourage log files to be archived by decreasing the logfilsz database configuration parameter:

db2 "update db cfg for sample using logfilsz 200"

The value 200 is the umber of 4K pages for each log file specified by logprimary and logsecond parameters.

10. Since the user exit database configuration parameter is set to YES, a full database backup is required before a connection to the database can be made. Make a note of the timestamp produced from the backup command as this wil be used during the database restore.

db2 "backup db sample to /home/db2v8_32/backups"

At this point the user exit program is enabled and ready to be used. The user exit program will be invoked every 5 minutes to check the active log directory for log files that need to be archived. Note that the larger the log file specified wit the logfilsz database configuration parameter, the longer it will take to fill a log file thereby making that log file more vulnerable to disk failure, corruption, etc. Depending on your transaction loads, you should only specify a log file size that will fill up in a reasonable period of time so that the user exit program can archive the log file off to a safe directory/location.

It is best to archive log files to a different disk at a different location. This adds a second level of protection.

<u>Testing</u>

Examples:-

Promote Archival Log Files:-

db2 connect to sample

db2 "insert into staff (select * form staff)"

db2 connect reset

(all database connections must bee released s that the database frees up resources and closes the log files)

Check the /mylogs/sample/NODE0000 directory.

The above scenario will truncate the currently active log file and copy it to the
mylogs directory.
You can test your logs by filling them with data. The database manager
Will check for full log files and copy them off to the archive directory.

Db2 "connect to sample"
Db2 "insert into staff (select * from staff)"

Remember that we updated the database configuration parameter logfilsz to 200 4K
pages (800K) in order to fill the log quickly.

Check the /mylogs/ directory.

If the user exit program is working correctly you should see an ARCHIVE.LG
created under /mylogs/ and one or more log files should be copied to /mylogs/
SAMPLE/NODE0000/

Now perform a database restore

The restore will invoke the user exit program to retrieve the log files from /mylogs/
directory back into the active log directory where they can be applied to the
database during roll forward recovery.

db2 connect reset
db2 "restore database sample from /backups taken at 20040416114420"
SQL2539W Warning!! Restoring to an existing database that is the same as the
backup image database. The database files will be deleted/ Continue? Y/N

The value 20040416114420 in the restore command is the timestamp taken from the
previous database backup. Since we are restoring on top of our existing sample
database we will receive the warning message above. Select Y and press enter.

Db2 "rollforward db sample to end of logs and stop"

Rollforward status

Input database alias = sample
Number off nodes have returned status = 1

Node number = 0
Rollforward status = not pending

Next Log file to be read =
Log files processed = S0000000.LOGto S0000004.LOG
Last committed transaction = 203-04-16-17.33.49.000000
DB2 Rollforward command completed successfully.

Placement of the ARCHIVE LOG file

A message will be written in /mylogs/RETRIEVE.LOG file forr each log file passed
back to the active log directory.

Parameter Count = 8
Database Name = Sample
Logfile name:_ S0000004.LOG
Logfilepath:-/SAMPLE/NODE0000/SQL00002/SQLOGDIR/
Node Number: NODE0000
Operating System:- AIX/UNIX
Request:- RETRIEVE
System Action:- RETRIEVE to
        /SAMPLE/NODE0000/SQL00002/SQLOGDIR/
        file S0000004.LOG from /mylogs/SAMPLE
Media:- Disk
User Exit Return Code: 0

The above is a basic introduction to using user exit programs with DB2 UDB. Note,
a user exit can be modified to suit your needs.

Index

A

About the Author

The author is a practicing database Consultant who specializes in DB2 and DB2 UDB. Having started as a Systems Programmer and then mastering IMS before moving on to DB2, the author brings over twenty years of broad and deep experience to the table. In addition the author has worked in twenty different countries and is considered a lead talent in the field.

The author has actually worked in Development, Production and Training, consequently his experience covers a wide arc of DB2.

This book does not pose as a complete guide to DB2, but rather, focuses on the important aspects of DB2 that would help end users, developers and database administrators alike.